Arthur Brown Willmott

The Mineral Wealth of Canada

A Guide for Students of Economic Geology

Arthur Brown Willmott

The Mineral Wealth of Canada
A Guide for Students of Economic Geology

ISBN/EAN: 9783337185862

Printed in Europe, USA, Canada, Australia, Japan

Cover: Foto ©berggeist007 / pixelio.de

More available books at **www.hansebooks.com**

THE MINERAL WEALTH

OF CANADA.

A GUIDE

FOR STUDENTS OF ECONOMIC GEOLOGY.

BY

ARTHUR B. WILLMOTT, M.A., B.Sc.

Professor of Natural Science, McMaster University; formerly Assistant in Mineralogy, Harvard University.

TORONTO:
WILLIAM BRIGGS,
WESLEY BUILDINGS.

C. W. COATES, MONTREAL.   S. F. HUESTIS, HALIFAX.

1897.

PREFACE.

For several years the author of this book has been giving a short course of lectures to his class in geology on the economic minerals of Canada. While it is not customary to treat this subject so fully in an elementary class, he has felt that in a young undeveloped country like our own, it was highly desirable that all university students should know something of our latent mineral wealth. So, at the expense of Palæontology, much of which is more suitable for an advanced course, time was found for economic geology in the elementary one.

To save the labor of dictation, and to make them useful to a larger number, these lecture notes are now published. They have been somewhat extended, to make the subject clearer to the general reader, who has not had any preliminary training in geology. So far as known, it is the only work giving a systematic account of the mineral resources of the Dominion. Originality, except in method of treatment, is not claimed. The work is a compilation founded largely on the excellent reports of the Geological Survey of Canada. These bulky volumes and the detailed statements in the reports of the Provincial departments of mines, while well and favorably known to the specialist, are almost unknown to the general reader, and unsuited for the elementary student. It is hoped that this book will not only prove serviceable itself, but that by its numerous references it will stimulate students to seek fuller information in the reports mentioned.

It has not been thought necessary in a book of this kind to burden it with references to the author whose work has been used. For the most part these works have been cited in the literature at the end of each chapter, but only those books appear which are likely to prove accessible to the student. Special works not usually found in small libraries have been omitted. Some changes have been made in the spelling of chemical terms, as recommended by the Chemical Section of the American Association for the Advancement of Science, and as adopted by the "Standard" Dictionary.

The kind assistance of several friends is gratefully acknowledged. To Dr. Coleman of the School of Practical Science, and to Mr. A. Blue, Director of the Bureau of Mines, the author is particularly indebted. The latter has read the work in proof, and special thanks are due to him for many valuable emendations.

TORONTO, August 10th, 1897.

ANALYSIS OF CONTENTS.

Chapter I.

Introduction . . . . PAGE 7

Comparison of the mineral resources of Canada with those of other countries—Description of rock-forming minerals—Origin of rocks—Kinds of rocks—Relative ages of rocks—Chart of geological time—General literature.

Section I.—MINERALS YIELDING METALS.

Chapter II.

Ore Deposits 21

Definition of ore—Usual combinations of the metals—Classification of ore deposits—Fissure veins—The filling of veins—Surface appearance of ores—Distribution—Erroneous ideas.

Chapter III.

Iron, Manganese and Chromium 40

Ores of iron—Impurities—Canadian localities—Production—Literature—Manganese—Chromium.

Chapter IV.

Nickel and Cobalt 50

Ores—Distribution—Geological occurrence—Uses—Production—Literature.

iv   CONTENTS.

Chapter V.

COPPER AND SULFUR 55
    Ores of copper—Geological occurrence—Canadian localities—History of mining operations—Production in Canada and other countries—Occurrence of sulfur—Uses—Localities where mined.

Chapter VI.

GOLD AND PLATINUM 66
    Comparison of Canada with other countries—Origin—Geological occurrence—Methods of milling—Canadian mines—Production.

Chapter VII.

SILVER, LEAD AND ZINC 81
    The ores of silver—Silver mines of Ontario and British Columbia—Production—Lead ores—Canadian mines—Zinc ores—Literature.

Chapter VIII.

ARSENIC, ANTIMONY, TIN, ALUMINUM AND MERCURY . . 92
    Ores of arsenic—Production in Ontario—Ores of antimony—Mines of New Brunswick—Ores of tin—Ores of aluminum—Occurrence of mercury in Canada.

Section II.—MINERALS YIELDING NON-METALLIC PRODUCTS.

Chapter IX.

SALT, GYPSUM AND BARITE 98
    Occurrence of salt—Origin—Localities in Canada—Manufacture—Production—Localities and production of gypsum and barite.

CONTENTS. v

CHAPTER X.

APATITE AND MICA 112
   Geological occurrence and production of apatite—Use
   —Occurrence of mica—Use and production.

CHAPTER XI.

ASBESTOS, ACTINOLITE AND TALC 119
   Composition of the minerals—Occurrence in Quebec—
   Method of Mining—Uses—Production—Literature.

CHAPTER XII.

PEAT, COAL, GRAPHITE 124
   Origin of peat — Uses — Localities — Kinds of coal —
   Analyses of a number of Canadian coals—Impurities in
   coal — Geological relations of coal — Origin of coal —
   Tables showing gradual passage from wood—Description of the different coal-fields—Production—Literature
   —Description of graphite—Occurrence — Use.

CHAPTER XIII.

THE HYDROCARBONS 148
   Composition of petroleum — Geological occurrence—
   Canadian oil-fields—Refining and use—Production—
   Composition of natural gas—Occurrence in Canada
   —Use and production — Asphalt — Anthraxolite —
   Albertite.

SECTION III.—ROCKS AND THEIR PRODUCTS.

CHAPTER XIV.

GRANITE AND SANDSTONE 161
   Uses of stone—Qualities of building stones—Production
   of granite—Origin of sandstone—Occurrence and use as
   building stone—Other uses of sand and sandstone.

CONTENTS.

CHAPTER XV.

CLAY AND SLATE 171
  Origin and composition of clay—Uses—Production—
  Origin of slate—Occurrence.

CHAPTER XVI.

LIMESTONE 179
  Origin and occurrence of limestone—Use for building
  material — Marble — Lithographic stone — Mortar and
  cement.

CHAPTER XVII.

SOILS AND MINERAL FERTILIZERS 187
  Origin of soil—Conditions of fertility—Ashes of plants—
  Analyses of some Canadian soils—Geological fertilizers.

APPENDIX 199
  Summary of mineral production, 1894 and 1895 —
  Tabular comparison of Canada with other countries in
  mineral production.

THE
MINERAL WEALTH OF CANADA.

CHAPTER I.

INTRODUCTION.

IN estimating the natural resources of our Dominion one thinks first of the boundless acres of fertile soil. These, a perennial source of wealth, which under good management can never be exhausted, are certainly our principal asset. At the same time it must be remembered that the annual production of both our forests and our fisheries amounts to many million dollars. Until recently the product of our mines was the least of these four resources, and this was not because we were without mineral resources, but that we had barely begun to exploit them.

Timber, fish, minerals are supplies laid up for us by Nature on which we can draw at will. Minerals once mined are never replaced. Timber once cut might be, but with us, never is, restored. Our fisheries we make some poor attempts to preserve. In agriculture alone do we seek to keep our rich inheritance intact. But though our mineral wealth be a fleeting one—though it be a resource which cannot be cultivated and increased like timber or fish—it is an asset of such enormous extent that it may be drawn on for hundreds of years to an amount far in excess of that annually produced by either our forests or our fisheries.

In considering the possibilities of mineral development, attention must first be directed to the extent and character of our country. With an area a little larger than that of the United States and with the same physical features, it would be strange indeed if much of the mineral wealth of that country were not duplicated north of the boundary. The Rocky Mountains and parallel ranges extend for some 1,300 miles through the States of New Mexico, Colorado, Wyoming and Montana, and for an equal distance through British Columbia and the Yukon District, and it is safe to assert that their mineral wealth does not stop at the forty-ninth parallel. So also the Sierra Nevada of California is represented north of the boundary by the Coast Range of British Columbia, and the latter may yet prove as rich as the former.

In the east the Appalachian system is perhaps even richer north of the boundary than south of it, though it is, of course, of much less extent. In the V-shaped territory of Archæan rocks stretching on either side of Hudson Bay from the Arctic to the St. Lawrence, there is an immense depository for minerals unequalled south of the line. True, we miss on the north the immense coal deposits of the Mississippi basin, but in a measure we have compensation in very fair-sized coal beds on both our Atlantic and Pacific coasts. It has been customary for Canadians to lament the existence of this large area of non-agricultural territory. But Nature always makes compensation. If by mountain upturning or glacial erosion she has rendered parts of our country unsuited for farming, she has in most instances at the same time raised and uncovered inexhaustible stores of silver and gold, of copper and iron.

Nearly the equal of Europe in size, we surpass any one nation of that continent in the variety of our mineral deposits, and may yet equal the richest of them in the total value of our production. Great Britain has had large deposits of coal, and her production is the greatest in the world. Her output must, however, shortly begin to lessen, while ours will increase. Russia stands second as a petroleum producer, and will no doubt surpass us for years. It is possible, however, that fields will be discovered in the North-West quite the equal of hers. The copper output of Spain at present exceeds ours, but the deposits here are quite as extensive as there. Similarly with other minerals, different European nations surpass us in production, but it is probable that our deposits are the more extensive, except in the case of coal, petroleum and tin. Already in asbestos we have surpassed not only Europe but the world. Italy, our only competitor, is far behind. With nickel we occupy the same proud position. Our gold product, though it may never equal that of Australia or the United States, may easily exceed that of all Europe combined.

Our deposits of iron, lead, silver, copper, salt and

other minerals are enormous. They are, however, almost entirely undeveloped. We can only guess at their value. So far we have, as a people, merely scratched the surface of a few acres of our mineral inheritance. Australia, with an area and population both slightly less than our own, has an annual mineral production nearly three times the value of ours. Belgium, a country of only 6,200,000 inhabitants, crowded into an area about half the size of Nova Scotia, draws twice as large an income from her mines as does Canada. And yet it is very probable that there is as much mineral wealth in Nova Scotia alone as in Belgium. Indeed, Nova Scotia, with coal and iron deposits in close proximity to each other and to the ocean, should, like Belgium, send her iron manufactures to the ends of the world.

While we have been slow in beginning the development of our mines a fair start has now been made, and we may hope for more rapid advancement in the near future. The total value of the mineral product for 1896 was about twenty-three and a half million dollars. Coal is the most important, yielding annually about eight million dollars. Gold is second, with a product approaching three million in value, which gives us tenth place among the nations. Nickel, copper and petroleum each exceed one million in value, and the silver output now amounts to over two million. In coal we rank eleventh, in petroleum fourth, and in silver tenth. Bricks and building stones are the only other products passing the million line in value. In ten years the total production has doubled. (See Appendix.) Within the last two years the gold and silver output of British Columbia has increased enormously. Estimated at $380,000 in 1893, it grew to about $2,200,000 in 1895, and reached $3,900,000 in 1896.

In succeeding chapters there will be given a description of the different economic minerals, the localities where they are found, and their uses and value. To do so will require the use of some geological terms, which we will now consider.

Rock-forming Minerals.—A mineral is an inorganic, homogeneous substance of definite, chemical composition. It may be a chemical element, more usually it is a compound resulting from the union of two or more elements in a definite proportion. A rock on the contrary is composed "of one or more simple minerals having usually a variable chemical composition, with no necessarily symmetrical, external form, and ranging in cohesion from mere loose débris up to the most compact stone." For example, granite is a rock composed of a variable mixture of the minerals, quartz, felspar and mica. Sandstone, limestone, sand and gravel are other examples of rocks. Gypsum is a mineral of definite composition, which in large masses may be considered a rock.

Minerals which are of economic value will be described later under the substance they yield. A brief description of the chief rock-forming minerals will be given here.

Quartz is the most widely disseminated mineral. It is readily distinguished by its glassy lustre and great hardness. It will easily scratch glass and cannot be scratched by a knife. It never breaks in flat surfaces but always in curved ones. In color it is usually transparent or white, though often stained yellow or red by iron oxid.

Felspar embraces several species which are much alike in physical features. All split in two directions with flat shining surfaces. In one variety, orthoclase, these cleavages are at right angles. In the other varieties, known collectively as plagioclase, they are nearly at right angles. The latter are sodium, calcium, aluminum silicates ; the former has potassium in place of sodium and calcium. The felspars can just be scratched with a knife.

The micas are easily known by their cleavage into thin elastic leaves. Some are clear and transparent, others black and opaque.

Pyroxene and hornblende are almost alike in composition but differ in their angles of cleavage. This is a distinction not evident in hand specimens of rocks. Both, as found in rocks, are dark green or black minerals with a hardness a little less than felspar. With a blowpipe they are much more easily fused.

Calcite is easily recognized when crystallized by the rhombohedrons or twisted cubes into which it readily breaks. All varieties are easily cut with a knife, and effervesce readily when touched with a drop of acid. In color calcite is usually white or grey. Dolomite differs from calcite in having magnesium carbonate mixed with the calcium carbonate of the latter. It effervesces with acids only when heated.

Chlorite occurs in thin leaves like the micas, but unlike them is not elastic. It varies in color from light to dark green. It is comparatively soft, and frequently has a pearly lustre.

Serpentine is usually a massive mineral with an oily green color and greasy feel. It is easily scratched with a knife. The fibrous variety is the asbestos of commerce.

Origin of Rocks.—The minerals described above with the occasional addition of a few others in subordinate amounts compose the bulk of our rocks. These constituent minerals are sometimes found with a more or less perfect crystal form, at other times with the edges rounded and worn. The particles vary in both cases from grains of microscopic size to masses of considerable dimensions. The rounded grains are evidently the result of moving water grinding down previously existing rocks. Rocks with this class of material are found to be arranged in layers as though due to beds of sediment deposited one on the other. These constitute the first great division of rocks known as the Sedimentary, Stratified or Fragmental Rocks. The second division embraces the Massive, Igneous or Eruptive Rocks, which have evidently solidified from a fluid condition either within the crust of the earth or after eruption from a volcano. The sharp angles of the crystals are preserved, and one mineral interlocks with another. These rocks present no appearance of bedding. The third and last division is known as the Schistose Rocks. They present characters intermediate to the other two. They are distinctly bedded, but do not show fragmental grains. The crystalline character of the constituents points to solidification from a fluid. In some cases they are doubtless sediments which have been subjected to sufficient heat to permit of the recrystallization of the minerals without destroying the stratification. For this reason they are often called the Metamorphic Rocks. In other cases they are Igneous Rocks, in which the divisional planes have been produced after the first consolidation.

Description of Rocks.—A few of the more important representatives of the above divisions will be described here.

Sand is an unconsolidated mass of fine worn grains of the harder minerals. Quartz is much the largest constituent since it resists decay, whilst the other minerals of the rocks, which are being worn down, are slowly carried off. Magnetite, an oxid of iron, is frequently abundant and gives a black color to the sand. Gravel is coarse sand.

Sandstone is simply consolidated sand, in some cases produced by pressure alone, in others due to a cementing material. The cement may be clay, iron oxid, silica, or calcite. The first gives rise to a clayey or argillaceous sandstone, which may graduate into a sandy or arenaceous shale. The red and yellow sandstones are due to oxids of iron.

A Conglomerate is formed of rounded pebbles up to a foot or more in diameter consolidated in any way. It bears the same relation to gravel and shingle that sandstone does to sand.

Clay results from the decay of felspars and similar silicates of the crystalline rocks. Deposited in water in beds it becomes more or less consolidated, and is then known as shale.

Limestones consist mainly of calcite or of calcite and dolomite. They also contain greater or less quantities of impurities—iron, giving them a red color; carbonaceous matter making them dark; clay, and silica or sand. They are usually grey or drab in color, of compact structure, and frequently contain organic remains. Some of them found associated with crystalline rocks have been metamorphosed by the action of heat and pressure, and are of a crystalline, granular texture. Fine-grained ones, susceptible of polish, are used as marble.

Granite is the most important of the massive or igneous rocks. It consists of an intimate mixture of quartz, felspar and mica. The crystals of these minerals may be barely visible or of considerable dimensions. The felspar may be red or white in color, and the granite is always of a corresponding hue. Granite occurs in masses of large extent and also in dikes in other rocks. Mica may be replaced by hornblende, the rock then being called a hornblende granite.

Felsite is an intimate mixture of exceedingly fine-grained felspar and quartz. It varies in color through grey, red and brown shades, is slightly translucent and can be fused with a blowpipe, while quartz, which it resembles, cannot.

Quartz-Porphyry.—Large distinct crystals of quartz or felspar are often found in felsite or in a fine-grained, microgranitic ground-mass. Such a rock is known as a porphyry.

Syenite is a granular crystalline mixture of orthoclase felspar and hornblende, usually red or grey in color. It differs from granite in the absence of quartz.

Diorite is a granular crystalline mixture of plagioclase felspar and hornblende. It is dark green to black in color, usually fine grained and often contains magnetite. Diabase, dolerite and basalt are closely related to diorite, and as all four weather to a green color they are often called greenstones.

Gneiss.—Among the schistose rocks gneiss is the most important. It resembles granite in being a crystalline mixture of quartz, felspar and mica. It has, however, a banded structure which seems in some cases to be the result of an earlier stratification. This laminated appearance is not always very distinct, and gneiss merges gradually into granite.

Mica Schist is a schistose aggregate of quartz and mica, each arranged in lenticular wavy laminæ. The mica may be the light or dark colored variety. Sericite mica may replace the ordinary micas, when a *sericite schist* results. Chlorite and talc with quartz and other minerals make respectively *chlorite schist* and *talc schist*. The last three are grey or green in color, with a pearly lustre and greasy feel. *Slate* results from the metamorphism and recrystallization in layers of ordinary clay and shale.

Relative Age of Rocks.—On examining any exposed section of the sedimentary rocks, it becomes at once evident that the older rocks are lowest in the series and the newer ones on top. In the same way it has been determined in many parts of the world that the sedimentary rocks rest on a fundamental complex of igneous rocks. In certain of the sedimentary strata coal seams are found in many parts of the world, and it at once becomes a matter of great interest to us as Canadians to know whether rocks of the same age occur here. Other strata are characterized by iron ores, or lead ores, and so on. Geologists have thus found it advantageous, from an economical as well as from a scientific standpoint, to correlate in age the various rocks of the world as far as possible. Three guiding principles are used:—1. That of superposition, that the newer rocks are above the older. In mountainous regions rocks have frequently been crumpled and overturned, and this principle cannot then be applied. Moreover, it does not help to correlate the ages of rocks not lying together. 2. The principle that rocks which are alike were formed at the same time. This is only true for limited areas, for, to take one example, sandstones formed ages apart are alike in composition and structure. 3. The principle that animal life was the same the world over at corresponding periods in the growth of each section of the sedimentary deposits. On studying the fossil remains entombed in the stratified rocks, it was found that certain formations contained trilobites in abundance, others graptolites, others fish, and so on. These characteristic animals were not confined to one horizon but were found in several. Beginning in one period they increased enormously in a second, and died out in a third. Other animal life, of course, existed along with them. The life of a period as presented to us in the rocks formed at the time, is thus quite sufficient to identify a rock formed at the same time in a remote part of the world.

In the study of English history it is customary to divide the subject into epochs. There is the Saxon epoch, the Norman epoch, the Plantagenet epoch, and so on. These are the great divisions, and under them are grouped the events which happened during the reigns of the successive sovereigns. Of course, the gradual development of the English nation went on irrespective of slight changes in rulers. But the reign of the sovereign, as the representative Englishman, makes a natural division of time. So in geological history, the development of animal types went steadily on, but the ascendancy of some particular group marks a division of time as does a dynasty in history. As to the relative lengths of the different geological time divisions little can be said. The main fact is the order of succession.

The oldest rocks are without fossil remains, and are called the Azoic or Archæan series of rocks, and are said to have been formed in Archæan time. Above these rocks are found the Palæozoic series; on these the Mesozoic series; on these again the Cenozoic series, which includes rocks now forming. These large divisions of time are subdivided as shown in the following chart, the oldest rocks being at the bottom of the page. The terms "time," "era," "period," "epoch," are divisions of time; the corresponding terms "series," "system," "group," "formation," refer to the rocks made during the interval of time. The first two divisions are of world-wide application; the latter are only of local use. The capital letters are those used on the Geological Survey maps for the respective formations against which they are placed.

CHART OF GEOLOGICAL TIME.

Time.	Era or System.	Period or Group.	Epoch or Formation.
Cenozoic.	Quaternary or Post-Tertiary, M.	Recent or Post-Glacial, M3. Glacial or Pleistocene, M1.	
	Tertiary, L.	Pliocene, L3. Miocene, L2. Oligocene, Eocene, } L1.	
Mesozoic.	Cretaceous, K.	Cretaceous, K.	
	Jurassic, J.	Jurassic, J.	
	Triassic, H.	Triassic, H.	
Palæozoic.	Carbonic, G.	Permian, G4. Carboniferous. Subcarboniferous, G1.	{ Coal Measures, G3. Millstone Grit, G2.
	Devonian, F.	Upper Devonian, F3. Middle Devonian, F2. Lower Devonian, F1.	{ Chemung. Portage. Hamilton. { Corniferous. Oriskany.
	Silurian, E.	Lower Helderberg, E6. Onondaga, E5. Niagara.	{ Guelph, E4. Niagara, E3. Clinton, E2. Medina, E1.
	Cambro-Silurian or Lower Silurian, D.	Trenton. Canadian or Quebec.	{ Hudson, D4. Utica, D3. Trenton, D2. { Chazy. Calciferous.
	Cambrian, C.	Upper Cambrian or Potsdam. Middle Cambrian or Acadian. Lower Cambrian or Georgian.	
Azoic or Archæan.	Huronian, B.	Upper Huronian. Lower "	
	Laurentian, A.	Upper Laurentian. Lower "	

20 THE MINERAL WEALTH OF CANADA.

LITERATURE.—Much excellent information on the economic minerals of the Dominion is to be found in the annual reports of the Geological Survey of Canada. Part " S " of the reports is issued separately, and deals entirely with the mineral production of the year. Geological maps of many areas are issued by the Geological Survey, and may be had for a few cents. A catalogue of the publications of the Survey will be sent on application to the Librarian of the Geological Survey, Ottawa. The reports issued yearly by the departments of mines of the provinces of Nova Scotia, Ontario and British Columbia are of great value. The Canadian Mining Review and the Canadian Mining Manual contain valuable summaries of particular industries, as well as many details of operations. The transactions of several of the Mining Engineers' Societies contain papers on Canadian mines.

For the characteristics of minerals and rocks the student will do well to consult Dana's "Manual of Mineralogy and Petrography." On the geological divisions of time see any good text-book, as Dana's "Manual of Geology," or Geikie's "Text-book of Geology "; also "Report of Geological Survey, Canada," 1882-84, p. 47.

SECTION I.
MINERALS YIELDING METALS.
CHAPTER II.
ORE DEPOSITS.

VERY few of our useful metals occur in nature as we employ them; nearly all are found combined with various elements to form chemical compounds. Sulfur, oxygen and carbonic acid are the chief mineralizers. Silica, arsenic, antimony and chlorin are also found united with the metals. These definite chemical compounds are called minerals. A mineral occurring in sufficient amount to be an economical source of a metal is called an ore. Associated with the metalliferous mineral there are usually others which constitute the gangue or vein-stone. This mixture of minerals makes the ore deposit.

Gold and platinum are nearly always found free and uncombined. Sometimes they are mixed with other elements to form alloys, gold frequently containing a percentage of silver, and platinum of iridium. Copper, silver and mercury are also found native at times, though more usually combined. Most of the metals form compounds with sulfur. Iron unites with it in two different proportions, but though widely spread neither pyrite nor pyrrhotite can be considered an ore of iron. Silver sulfid, or argentite, is an important ore of silver. So also are several sulfids of silver and antimony, and silver and arsenic. Cinnabar, the sulfid of mercury, galena, the sulfid of lead, stibnite, the sulfid of antimony, are the main sources of these metals. Zinc sulfid or blende, and chalcopyrite, bornite and chalcocite, three copper sulfids, are important ores of these two metals.

The oxids of iron, manganese and tin constitute the most important ores of these metals. Oxids of copper, and of zinc, are also extensively mined. Among important carbonates are those of iron, copper, zinc and lead. Silicates are not often a source of metals, but calamine, chrysocolla and garnierite are mined respectively for zinc, copper and nickel. Cerargyrite, or silver chlorid, is the only chlorid of economic importance. Arsenopyrite, a compound of arsenic, iron and sulfur, frequently carries gold. Arsenic also unites with nickel, and with cobalt, to form ores of these metals.

Several of these minerals are often closely associated. Silver and lead sulfids are so frequently mixed that it hardly pays to mine lead ore unless it contains some silver. Silver and zinc sulfids are also frequently associated. Iron and copper pyrites are often intermingled; so also, iron and manganese oxids. Gold is commonly associated with iron or copper pyrites, though these may have been oxidized on the surface of the deposit.

Other minerals of no economic value are usually associated with those mentioned above. The most common of these gangues are quartz, calcite, barite and fluorite. Sometimes, as in the iron deposits, the gangue is relatively small; in most cases it constitutes the great bulk of the deposit. If one-twentieth of one per cent. of a gold deposit were gold, *i.e.*, about a pound in a ton, the ore would yield $300 to the ton, while $20 would in most cases be very profitable. Evidently in deposits of the precious metals the ore is a minor accessory. In all cases the deposit must be concentrated—the vein-stone must be separated. This is usually accomplished by currents of water which carry off the light gangue and leave the heavy mineral.

Ore deposits are the result of the concentration of mineral particles once widely disseminated through the surface rocks or too deeply seated to be of use to man. They may consequently be classified according to the manner in which they were formed. Unfortunately our knowledge of their origin is far from perfect, and most authors adopt an empirical classification based on the form of the deposit. This has its advantages, since it appeals to the practical man who is more concerned about the form and permanence of his deposit than about the origin. Many schemes have been proposed. That of Louis ("A Treatise on Ore Deposits." Phillips and Louis, 1896) is among the best, and will be followed here:

CLASS I.—*Symphytic Deposits, or those formed at the same time as the enclosing rocks.*

(*a*) Clastic deposits.
(*b*) Precipitates from aqueous solution.

(c) Deposits from solution subsequently metamorphosed.

(d) Disseminations through sedimentary beds.

CLASS II.— *Epactic. Deposits, or those formed subsequently to the enclosing rocks.*

Sub-class 1. Veins:

(e) Fissure veins.

(f) Bedded veins.

(g) Contact veins.

(h) Gash veins.

Sub-class 2. Masses:

(i) Stockworks.

(j) Massive deposits in limestone.

(k) Massive deposits connected with igneous rocks.

(l) Disseminations in igneous rocks.

Symphytic Deposits.—These have been laid down as beds in sedimentary rocks and have subsequently been subject to the same folding as the enclosing sediments. They may now be found in synclinals or basins, or in anticlinals or saddles. These ore deposits, like all other sediments, may be affected by fissures and faults. Portions of a bed originally continuous may thus be found at very different levels on opposite sides of a fissure. The fault may also cause a horizontal separation of hundreds of feet. When the fault is vertical no horizontal displacement occurs. More frequently the fault is inclined, and dislocation results according to the following law: The portion of the bed that lies on the inclined plane slips

THE MINERAL WEALTH OF CANADA. 25

down relatively to the other part. Or, as it is stated for the miner, "if in driving on a bed a fault is met with in the roof, go down; if first in the floor, go up, to find the faulted portion."

(a) The clastic deposits have been produced by the disintegration of more ancient metalliferous deposits. This may have occurred at the present position of the ore, but usually water has transported and assorted the products of decay. The black iron sands, magnetite and ilmenite, are the most wide-spread representatives of this class in Canada. Along the Great Lakes and especially along the Lower St. Lawrence, immense bodies of these sands are met. They are due to the decomposition of the basic rocks of the Laurentian. Owing to their high percentage of titanium they are of little value as a source of iron. More important from the economical standpoint are the auriferous gravels of British Columbia and the sands of the Chaudière, Quebec. The heavy gold brought from the mountains by the streams was deposited on the current being checked These irregular beds are known as placers. The process has been going on in all geological periods, and auriferous gravels are known which were formed by rivers in Cambrian times. Platinum is entirely derived from similar placers. Tin, in the form of the oxid, is also largely won from river gravels.

(b) The ores of iron and manganese are practically the only ones formed by precipitation from aqueous solution. The process has taken place in all ages and is still at work. The acids resulting from the decay of plant life are good solvents of the oxids of iron so widely distributed in the igneous rocks. The carbonate of iron found in some limestones is soluble in water impregnated with carbonic acid. Iron pyrite oxidizes to ferrous, or ferric sulfate, both soluble salts. In these ways great quantities of iron are leached from the rocks and carried to ponds, where, exposed to the action of the air, carbonic acid is evolved and the iron precipitated either as the carbonate or as the hydrated oxid. Limonite, or bog iron ore, is essentially the hydrated peroxid of iron (Fe_2O_3 + 3 H_2O), though impurities are often present. There is no doubt but that it is formed in the way indicated. This ore is found quite extensively near Three Rivers, Que. It occurs in swamps one to fifteen feet below the surface in patches from three to thirty inches thick, and from a few square feet to several acres in extent. Similar ore is found in lakes in Quebec and Sweden. The deposits are dredged, and it is found that they are renewed quite rapidly. In ten to twenty-five years economic amounts have been known to form. Clay iron-stone, or argillaceous carbonate of iron, is found in the Carboniferous rocks of Nova Scotia. It has doubtless been formed in the same way as the more recent deposits.

(c) The deposits of this group were probably formed just as those of the previous one, but were afterwards subjected to metamorphism. The oxids of iron, hematite (Fe_2O_3), and magnetite (Fe_3O_4), are the great representatives of the group. These ores were probably deposited as the hydrated oxid in swamps or

THE MINERAL WEALTH OF CANADA. 27

lakes. Subsequently the bog ore was covered by sediment, and the whole subjected to heat and pressure. The water was driven from the ore and the materials of the sediment recrystallized. In many cases the beds were upturned, and the present ores seem at times to be in veins rather than in beds. For the most part they occur in rocks of Laurentian, Huronian and Cambrian age. Scores of examples are afforded by the Archæan of Canada.

(d) The ores disseminated through beds form a very important group economically. Genetically they connect the two great classes of ore deposits. The main mass of the rock, the non-metallic portion of the deposit, is of sedimentary origin. The metallic portion was introduced later, probably in solution. Some have held that the metallic portion also is of sedimentary origin. We know, however, of no process by which lead sulfid, copper sulfid or gold may be precipitated from sea-water. On the contrary, we do know that, under certain circumstances, subterranean water may carry these materials in solution. Indeed, it is in this way that fissures have been filled. Two examples of dissemination may be mentioned. In the Permian rocks of Mansfeld, Germany, there is a shale impregnated with several copper minerals, which has been mined for centuries. The bed, which is only a foot and a half thick, extends for miles. The rich gold deposits of the Witwatersrand, South Africa, are of similar origin. Sand and conglomerate beds, quite destitute of gold, were here upturned and faulted. Concurrently subterranean waters bearing gold in solution penetrated the more porous beds. The conglomerates thus contain most of the gold—the sandstones but little.

Epactic Deposits.—All the ore deposits of this class were formed subsequently to the enclosing rocks, consequently fragments of these rocks are often found in the ore body. With the exception of iron the larger proportion of every metal is derived from this class of deposits. Two subdivisions of the class are recognized depending on the form of the deposit. Under the term vein is included the tabular deposits, which have considerable length and depth but small breadth. The mass deposits include the remaining irregular ones, which have no definite shape and are of varying size.

(e) Fissure veins have originated in dislocations of the country rock, caused by movements of the earth's crust; subsequently they have been filled with mineral matter. A dike, which bears a superficial resemblance to a fissure vein, differs in that it has been formed by an intrusive sheet of igneous rock. Its constituents are generally non-metallic. A true fissure vein cuts across the planes of bedding of a sedimentary rock.

The walls of a vein are seldom parallel for any distance. This is due to the fact that there has usually been a slipping or faulting along the fissure. Conceive an irregular crack in the crust, and that one side has slipped downwards, and the walls will no longer be parallel; on the contrary there will be a succession of narrow and wide parts of the vein, if, indeed, it does not pinch out entirely at places. Connected with this movement there will be a grinding of the two walls, which often leaves a peculiar smooth surface, with parallel scratches called slickensides. A fine powder also results. This with water forms a seam of clay—the selvage of the vein. Most of these fissures are vertical or nearly so. The greatest angle of inclination which they make with the horizon is called their dip. The horizontal direction at right angles to this is called the strike. With inclined veins the upper wall is known as the hanging wall; the lower as the foot wall.

In size veins vary greatly. Some have been traced for several miles in length; others have been mined to a depth of half a mile. In thickness they vary from a minute crack to many yards. From their mode of formation they are believed to extend indefinitely in depth. Because of their persistency and regularity, true fissure veins are looked on with most favor by the miner.

The ultimate cause of the formation of fissures is probably to be found in the cooling of the earth's interior. As this portion of the globe cools it must contract, and this necessitates the folding in of the outer crust. This crust must be crumpled and folded to permit of its occupying less space, and fissures would naturally occur parallel to the axis of folding. The settling down of the upper rocks would produce forces of compression and torsion, and Daubrée has shown experimentally that in this way two sets of fissures, at right angles to each other, would be produced. This is in accordance with the facts noticed in many mining regions. Some fractures may be due to the contraction of a cooling mass of igneous rock; others are, perhaps, caused by the drying of a sedimentary rock, and consequent contraction and fissuring. Most fissures are, however, the result of dynamic causes, not of contraction.

The fissure being formed, it is next in order to inquire how it was filled. Before discussing this point certain characteristics of veins should be noted. As a usual thing the larger part of every vein is occupied by the non-metalliferous gangue. Quartz, calcite and fluorite are common vein-stones. They are crystalline in structure, and are often arranged in layers on the walls. The metallic portion of the vein is very irregularly distributed. In few cases does it pay to remove the whole of the vein-stone, and only the richer parts are hoisted to the surface. Sometimes the metallic portion is concentrated in a horizontal band in the vein. This is known as a course of ore. At other times the metal-bearing minerals are concentrated in somewhat vertical bands in the plane of the vein. These are known as shoots (also written *chutes*) of ore, or chimneys. The shoots of a vein are usually parallel to one another, and the angle of inclination is most commonly that of the bedding or cleavage of the rocks in which the vein occurs. When the ore occurs in detached patches it is said to be bunchy.

The nature of the country rock seems to often exert great influence on the ore body. In Cumberland, England, it has been noticed that the veins enclosed in limestone, sandstone or schist are more productive than those between walls of slate. In Derbyshire the veins traverse igneous rocks and also shales and sandstones. In the latter the veins are productive; in the former the lead ore is usually absent. At the famous Silver Islet mine, Lake Superior, the ore was found in a vein intersecting a diabase dike in argillite. The vein was exceptionally rich in the diabase, but barren in the argillite. Depth has no known influence on the character of a vein.

The Filling of Veins.—After the formation of the fissure it was filled with gangue and ore. Where were the materials found, and how were they transported to the vein? Seven distinct theories are tabulated by Louis, some of which have only an historical value:

1. Theory of Contemporaneous Formation.
2. Theory of Electric Currents.
3. Theory of Aqueous Deposition from above.
4. Theory of Igneous Injection.
5. Theory of Sublimation.
6. Theory of Lateral Secretion.
7. Theory of Ascension.

The first three may be dismissed as incorrect. The fourth, while the acknowledged mode of formation of dikes of igneous rocks, does not account for many characteristics of veins. Sublimation probably accounts satisfactorily for the presence of mercury and cinnabar throughout a rock. The theory supposes the metal to be volatilized in the hot interior of the earth and deposited in the cool part of the vein above. It fails to account for the vein-stones, and so cannot be accepted for many deposits.

The theory of lateral secretion was put on a firm basis by the labors of Sandberger. He taught that water percolating through the country rock had, by means of natural solvents, such as carbonic acid, leached from it the materials which were afterwards deposited in the vein as the water evaporated. By careful chemical examinations he showed that all the common metals were to be found in the silicates of the crystalline rocks. Pyroxene, hornblende, the micas and the felspars were the depositories whence not only copper, lead, zinc, etc., were derived, but also the gangue materials, silica, fluorin, etc.

Sedimentary rocks, apart from the limestones, consist of the débris of the older crystalline rocks. Consequently the metal-bearing silicates, finely comminuted it may be, should also be present in stratified rocks like shale and slate. Lead, copper, zinc, arsenic and others were actually found in clay slates. Thus he proved that the metals occurred in rocks of every geological age.

This theory explains fairly well the origin of the metals and gangue, accounts for the frequent banded structure of a vein, explains the fact that shoots usually follow the dip of the enclosing rocks, and gives a good reason for the changes which take place when a vein passes from one formation to another. Against it may be urged that different sets of fissures traversing the same formation often contain very different ores. It is also to be noted that a vein traversing several formations often contains the same ore.

The theory of ascension had as its strongest supporter Posepny, of Germany. He believed that the vein material is carried in solution from the hot interior of the globe. Opposing the view that the metals are derived from the crystalline rocks, he supposed a heavy metalliferous layer at a considerable distance below the surface. Water slowly forcing its way down becomes superheated, and under the great pressure is an active solvent. In this way the metals and vein-stone are leached from the rock, carried into the vein and deposited above. Veins are actually being formed to-day in this way in Nevada and California. The theory avoids some of the difficulties of the previous one, but creates others.

American geologists are inclined to accept a theory combining the best points of the last two. Le Conte asserts that the source of the metals is a leaching of all the wall rocks, but mainly the lowest portions. Metals have been brought up by ascending currents, and smaller contributions have come from the upper rocks. Highly alkaline water was the main solvent. The sulfids were the chief minerals dissolved, and deposition took place in all kinds of fissures. The deposits are found mainly in mountainous regions and in metamorphic and igneous rocks, because there the fissures were made and the heated layer occurs nearest the surface.

A fissure vein has not always two well-marked walls.

Frequently one or both are wanting. The alkaline silicate in its upward passage in the fissure often attacked the wall rock, and exchange of molecules occurred. Parts of the rock were dissolved and carried off—some of the ore was deposited in its place. In this way the wall disappeared, and the vein was widened in an irregular manner.

(*f*) Bedded veins are parallel with the bedding or foliation of the country rock, while the previous class cut it in all directions. This class of fissures is due to a plane of weakness in the bedding, or to a folding of the beds which has left a cavity. They are not so continuous as true fissures, but one vein usually succeeds another. They vary considerably in thickness, and are often lenticular; many of them do not appear at the surface. They may be faulted like an ordinary fissure vein; the gangue and ore are alike in both classes. Gold particularly is found in bedded veins, those of Nova Scotia being good examples.

(*g*) Contact veins are cavities between dissimilar rocks which have been filled with ores through the influence of one of the rocks. Obviously they resemble bedded veins in appearance, except where one rock is eruptive. An excellent example is afforded by the deposits of Leadville, Col. Igneous dikes have here crossed beds of limestone. Mineral-bearing solutions passing up the line of weakness between the two rocks have dissolved the limestone and replaced it with silver-lead ores.

(*h*) Gash veins are properly irregular deposits made in the joints, and between the beds, of limestone.

They are of small extent, and do not pass vertically to any distance. Water, charged with carbonic acid, has probably dissolved the rock along the joint-plane, and subsequently mineral matter has been deposited from solution. The lead and zinc ores occurring in the Trenton limestone of Iowa and Missouri are the best examples.

(i) A stockwork consists of a mass of igneous, metamorphic or stratified rock, "impregnated with metalliferous mineral, either in the form of small reticulated veinlets, or more or less uniformly disseminated through the rock in connection with the veins." The mass has no definite limits, and merges gradually into the surrounding rock. Typical examples are the tin deposits of Saxony and of Cornwall. Apparently the rocks containing the tin ore have been shattered, and mineral-bearing solutions rising in the fissures have deposited their burden there or exchanged part of it for a portion of the wall rock. This group of deposits is accordingly related to the true fissure veins.

(j) Massive deposits in calcareous rocks seem to be due to the slow replacement of the soluble limestone by the ore of a mineral-bearing solution. Apart from their irregular form they closely resemble gash veins, and should perhaps include them. The deposits are very irregular in size and shape. Many of the silver deposits of Nevada afford good examples of this class.

(k) Masses in igneous rocks are either irregular or lenticular in shape, and are found either in the rocks or at the plane of contact between them and an older rock. They resemble somewhat contact veins, but are not tabular like them. Oxids of iron and sulfids of iron, of copper and of nickel are the chief minerals of this class of deposits. The sulfids have probably been introduced in solution in cavities which were subsequently enlarged by the exchange of the mineral for the rock. A typical example is afforded by the copper and nickel deposits of Sudbury, Ontario. Here the ore is found in lenticular masses, either in diorite or at the contact of the diorite and the Huronian schists which it pierces.

Immense deposits of magnetite and hematite are found in the Archæan rocks of Ontario and Quebec. They are irregular in shape, and occur in igneous rocks or crystalline limestone. By some authors they are classed here, though others assert that they are metamorphosed sediments and belong to group c.

(l) Disseminations in igneous rocks include (1) deposits resembling the last, but where the metalliferous part is so scattered that the whole rock must be removed; (2) deposits where an igneous rock has been impregnated rather than a stratified one, as in d. A typical example is afforded by the native copper deposits of the basin of Lake Superior.

Surface Appearance of Ore Deposits.—In most cases ore deposits are very different on the surface to what they are when opened. At a few feet below the surface, the distance varying with the locality, a zone of water, known as the water-line, is met. Above this, air, water and chemical agents may react on the ore, and the usual result is oxidation. Hydrates, carbonates, sulfates and chlorids may also be formed. Many of these are soluble and are carried off by water. These surface accumulations are called gossan. The French name, *chapeau de fer*, and the German, *eisen hut*, both meaning "iron hat," are very expressive. Iron pyrites is a very widely disseminated mineral, and on oxidation it yields the hydrated oxid, limonite, reddish to brown in color. In and beneath this layer there is often found a rich deposit of gold, silver or copper, as the case may be. The weathering of the vein has permitted the removal of the gangue and the concentration of the heavier metals. From this fact arises the German proverb:

> "A mine is ne'er so good as that
> Which goes beneath an iron hat."

Below this again the water-line is reached, and the character of the ore may change entirely. For instance, a gold ore may be free-milling on the surface, and below become most refractory. A case in point is afforded by the gold ores of Hastings, Ontario. Rich and free-milling on the surface, they rapidly became arsenical and rebellious. Lead and zinc may exist as the carbonates on the surface, and pass at the depth of a few feet into the sulfids, galena and blende.

Distribution of Ore Deposits.—A consideration of the methods of formation of ore deposits would lead us to expect them where one or more of the following conditions are presented: 1. A region of disturbance, where fissures may have been made and circulation promoted. 2. A region where heat has been at work. This may have been due to volcanic action or produced by metamorphism. 3. Where the solvent action of water has been enormously increased by the pressure of overlying rocks and by the greater heat. 4. Where action has been long continued, and feeble agencies may thus have been able to effect considerable change. In accordance with these conditions we find the great majority of ore deposits (1) near eruptive rocks, especially the earlier ones; (2) in mountainous regions, particularly those which have been well denuded, as shown by their low rounded forms; (3) in regions of ancient rocks.

Erroneous Ideas Regarding Ore Deposits.—1. It is often asserted that true fissure veins are likely to increase in width as the shaft is sunk. The truth is that they will widen and narrow alternately, sometimes pinching out entirely. If at the *present* surface a vein is narrow it may widen for a time; if, on the contrary, it is struck at a wide part it may narrow for a time. A good illustration of this, both as regards changes in the depth and the length of a vein, is a torn paper, with the parts slightly shifted to show the faulting.

2. Fissure veins are said to grow richer as depth increases. Apart from the enriching at the surface due to the decay and removal of the vein matter, this is hardly true. The ore in a vein is always irregularly distributed. In sinking the miner will, of course, pass from poor portions to richer ones and then on to lean ones again.

3. It is often held that certain directions of strike

in veins indicate rich or poor deposits. This can only be true of limited regions where the parallel fissures may be supposed to be due to the same cause. Those formed at the one time are likely to have been filled with the same solution. An earlier or later set of fissures might have been filled with a different solution containing no metallic ore, or a different one. The strike of veins containing the same ores may be widely different in different localities.

4. The country rock certainly exerts an influence on the vein material, and preference for a particular kind on the part of the miner is justifiable within limited regions. Nevertheless, a wall rock which is barren in one district may prove to be rich in another.

LITERATURE.—"A Treatise on Ore Deposits," Phillips and Louis, 1896. "The Genesis of Ore Deposits," Posepny, Trans. Am. Inst. Min. Eng. XXIII., 197-369. Newberry, "School of Mines Quart.," 1880, V. 337.

CHAPTER III.

IRON, MANGANESE AND CHROMIUM.

Ores of Iron.—Among the metals iron is easily of first importance, because so indispensable to all our industrial undertakings. It is widely distributed in nature, occurring as an oxid and as a carbonate. Magnetite (Fe_3O_4) is richest in metallic iron, containing 72 per cent. when pure. It can always be attracted by a magnet, and often is itself able to attract soft iron. It is with difficulty scratched by a knife, and yields a black powder. Some varieties contain manganese, others titanium. Hematite (Fe_2O_3) contains, when pure, 70 per cent. of iron. Several varieties are distinguished, all of which yield a dark reddish powder. The hard crystalline kind, with a steely lustre, is called specular ore; a black, shining, scaly ore is known as micaceous hematite. Mixed with clay it yields a brown-black to reddish colored ore of dull lustre. The harder mixtures are clay iron-stones; the softer are red ochres. Fossil ore consists of red oölitic grains. Part of the iron of hematite is often replaced by titanium. Brown hematite ore includes a number of minerals, all of which are hydrated oxids, such as limonite ($2Fe_2O_3 + 3H_2O$), göthite, etc. These minerals yield water when heated, give a brown powder and streak, and contain 60 per cent. or less of iron. Iron carbonate, called siderite or spathic iron ore, contains about 48 per cent. of iron. It is brown in color, cleaves readily into rhombohedrons, and effervesces when heated with acids. In coal regions it is frequently found mixed with earthy matter, and is then known as clay iron-stone. Mixed with bituminous matter, it forms black band.

Clay iron-stone, though containing a smaller amount of iron, is often more valuable than richer ores because of its proximity to coal and fluxes. Ores of iron are so widely distributed and in such large amounts that only those deposits which are favorably located can be utilized. The value of an iron deposit depends on (1) its proximity to fuels and fluxes needed for its reduction; (2) its freedom from injurious materials not readily removed in smelting; (3) the percentage of iron which the ore will yield.

Anthracite, coke and charcoal are the usual fuels. Limestone is the flux employed to remove the common impurities of clay and quartz. The proximity of these materials in Nova Scotia has caused a development of the iron industry there, while the rich ores of Ontario are neglected. Other impurities are phosphorus, sulfur and titanium. A small amount of sulfur causes an iron to be "red-short," that is, brittle and difficult to work at a red heat. One-tenth of one per cent. of phosphorus causes the metal to be "cold-short" or brittle when cold. Ores containing these elements are unsuited for the manufacture of steel. By lining the converter with a magnesium or calcium mineral it has been found to be possible to use many ores formerly rejected because of their phosphorus. Titanium does not injure the iron, but the presence of any amount in the ore increases the expense of reducing it.

Geological Occurrence.—The ores of iron are found particularly in the oldest formations. The Laurentian, Huronian and Cambrian are the great iron ages. The ores in rocks of these periods are hematites and magnetites, especially the latter. Hematites are found in Silurian and Devonian strata in Nova Scotia. Siderite is found in the Palæozoic of Nova Scotia, and in the form of clay iron-stone throughout the Cretaceous and early Tertiary of the North-West. Limonite is abundant in the Silurian and Devonian of Nova Scotia, and its modern representative, bog iron ore, is found in the Post-Tertiary of Quebec and Ontario. This last has been dissolved by organic acids from the crystalline rocks, and deposited in swamps after oxidation. The beds in the Archæan are doubtless metamorphosed bog ores, though in some cases they may be of an eruptive origin.

Canadian Localities.—*Maritime Provinces.*—Iron ores occur in large amounts in Nova Scotia. All varieties are represented, and are found in nearly every geological age. Active operations are confined to the counties of Pictou, Annapolis and Colchester. These counties respectively produced 31,000, 30,000 and 18,000 tons of ore in 1895. In Pictou the ores are found along the East River close to the coal field.

In Devonian strata beds of brown hematite with specular ore and siderite are found. An oölitic hematite resembling the Clinton ore of the United States occurs in Silurian beds. The largest deposits, and the only ones yet worked, are found at the contact of the Carboniferous rocks with earlier formations. The ore is mostly brown hematite. Two companies are mining and smelting these ores. A charcoal furnace is used at Bridgeville, and Bessemer pig is made with coke at Ferrona. At Torbrook, Annapolis county, there is a considerable area of hematite ores. The beds are three to twelve feet thick, and the ore is of good quality. It is shipped to Londonderry and Ferrona. At the Acadia Mines, Colchester county, there is "an extensive development of brown hematite in a vein in Devonian strata associated with specular ore, ochre, ankerite and other carbonates of lime, iron and magnesia." This ore, mixed with hematite from Torbrook, is smelted by the Londonderry Iron Company. The product is largely sold in Montreal.

In Cape Breton there are numerous beds of hematite and magnetite in Archæan strata. Specular ore is found in Guysboro' and hematite in Antigonish. Other localities are Pugwash, Grand Lake, Brookfield, Goschen, Selma, Clifton, etc.

In Carleton county, N.B., beds of hematite are found in Lower Silurian slates. A charcoal furnace was in blast for some time at Woodstock, and several thousand tons of iron made.

Ontario and Quebec.—Bog iron ores were first

44 THE MINERAL WEALTH OF CANADA.

discovered in the district of Three Rivers in 1667. The first forges were erected in 1733, and iron has been smelted in the district almost continuously since that date. The Radnor Forges near Three Rivers are the present representative of the old industry. Bog ore is procured on both sides of the St. Lawrence, and charcoal made in the vicinity is used for fuel. The product is particularly adapted for car wheels. At Drummondville, on the St. Francis, are two other furnaces also using bog ore and charcoal. The ore is mined partly in the vicinity and partly in Vaudreuil. Bog ores are quite abundant in the low lands flanking the Laurentian hills on the north of the St. Lawrence. In the Archæan rocks north of the Ottawa and St. Lawrence immense beds of magnetite and hematite are found. Below Quebec these often contain considerable titanium, but to the west many of them are excellent ores. Beds twenty-five feet wide are of common occurrence. For the most part they are interstratified with gneiss. In the metamorphic rocks of the Eastern Townships other important deposits are found. Except for the occasional export of a few tons these oxids are unused.

In Ontario similar beds of hematite and magnetite are found in Archæan rocks. Large amounts have been mined at several localities, but no regular operations are going on at present. Most of the ore was exported; some of it was smelted at furnaces now dismantled. The chief mining locations are along the Kingston and Pembroke Railway; in Hastings, Peterboro' and Victoria counties; north of Lake Huron;

west of Lake Superior. In the last district, on the Mattawin and Atikokan rivers, bodies of ore are found which resemble in appearance and mode of occurrence the famous deposits of Minnesota. The ores of Gunflint Lake are a continuation northwards of the Mesabi range of Minnesota.

Bog ores are found at a number of places in southwestern Ontario. They were smelted early in the century, and are again being mined for a new furnace at Hamilton. This furnace also uses hematite and magnetite from other parts of Ontario. Siderite is reported as occurring in large deposits in the Devonian, on Moose River.

Western Canada.—Clay iron-stone occurs at a number of places throughout the lignite Tertiary of the North-West, but nowhere in economic amounts. It is also found in the coal series of British Columbia. Magnetite is, however, the chief ore of this province. It has been mined at Kamloops Lake, Redonda Island and Texada Island for export. It is found in many localities and of good quality. The ore bed at Texada is twenty to twenty-five feet thick, and extends for a mile with a thickness of one to ten feet.

Production.—Canada is particularly backward in developing her iron industries. Few countries have larger deposits of ore, and much of it is convenient to coal and flux. The smallness of the market is the great difficulty. Moreover, Nova Scotia, the chief producer, is some distance from Ontario, the chief consumer. The following tables will give an idea of the industry :

THE MINERAL WEALTH OF CANADA.

Materials Made and Used.	1894.		1895.	
	Quantity.	Value.	Quantity.	Value.
Pig iron made—tons	50,000	$647,000	52,000	$696,000
Iron ore consumed—tons	109,000	224,000	93,000	218,000
Fuel con- sumed. Charcoal—bush.	1,174,000	54,000	790,000	32,000
Coke—tons	52,000	142,000	49,000	139,000
Coal—tons	8,000	15,000	3,000	5,0,0
Flux consumed—tons	35,000	34,000	32,000	30,000

By provinces the production of ore in 1895 was:

Nova Scotia.................... 83,792 tons.
Quebec......................... 17,783 "
British Columbia .... 1,222 "

Total...................... 102,797 "

In 1895 the exports of iron and steel goods amounted to $175,000 and the imports to $8,002,000. There was further imported scrap iron, etc., to the value of $697,000, and against this an export of ore valued at $4,000.

Compared with foreign countries the Canadian production is insignificant. The following table is compiled from Rothwell's "Mineral Industry":

PRODUCTION OF IRON AND STEEL IN THE WORLD, 1895.

Country.	Pig Iron.	Steel.	
United States	9,597,000	6,213,000	
Great Britain	8,022,000	3,150,000	
Germany	5,789,000	2,825,000	
France	2,006,000	717,000	
Russia	1,454,000	574,000	Metric tons of 2204 lbs.
Austria	1,075,000	495,000	
Belgium	829,000	456,000	
Sweden	465,000	230,000	
Spain	206,000	65,000	
Canada	38,000	.....	
All others	387,000	329,000	
	29,868,000	15,054,000	

THE MINERAL WEALTH OF CANADA. 47

Great Britain and Germany are relying more and more on imported ores. Spain, which ranks fourth as a producer of iron ore, exports considerable to Britain. Sweden also ships ore to that country.

LITERATURE.—History of manufacture in Canada, Bartlett, Trans. Am. Inst. Min. Eng. XIV. 508 ; Canadian Mining Manual, 1896. Theories of Origin, Phillips and Louis, "Ore Deposits"; Winchell, Bull. 6 Minn. Geol. Sur. Statistics, Rep. S Geol. Sur. Can. Localities, Catalogue of the Museum. Nova Scotia : Pictou, Geol. Sur. V. 1890, 175 P ; Trans. Am. Inst. Min. Eng. XIV. 54; Reports Dep. of Mines; "Acad. Geol." New Brunswick: Geol. Sur. 1874. Quebec: Geol. Sur. IV. 1888 K. Ontario: Geol. Sur. 1873-74; Bur. of Mines, 1892. British Columbia: Rep. Geol. Sur., III. 1887 R.

MANGANESE.

The ores of manganese are almost wholly oxids, or hydroxids, though the metal occurs in many other forms. It is similar to iron in its chemical affinities and geological distribution, so that it often occurs with ores of that metal. Pyrolusite (MnO_2), the dioxid of manganese, is the most important mineral by reason of its purity. Wad, or bog ore manganese, is more widely distributed, but is often useless through the presence of sulfur, phosphorus, etc. Psilomelane, manganite, braunite and hausmannite are other manganese minerals. Some silver and some zinc ores contain a considerable amount of manganese, which is saved as a by-product.

The great use of the metal is in the iron industry. Nine-tenths of the product is converted into spiegeleisen and ferro-manganese, two alloys with iron containing from one to ninety per cent. of manganese. These alloys are invaluable in the manufacture of steel. Not only does the manganese prevent the oxidation of the iron, but a small per cent. increases the strength of the steel. Because of the readiness with which pyrolusite yields oxygen, it is used in the manufacture of chlorin and as a decolorizer of glass. Compounds of manganese are also used as coloring materials in calico-printing, coloring glass and pottery, and in paints. For these chemical processes only the purest pyrolusite is available, whilst for spiegel-eisen an ore containing iron, water or calcite may be used.

Pyrolusite, manganite and wad are widely distributed through the Lower Carboniferous rocks around the Bay of Fundy. The first systematic mining operations were begun at Tenny Cape, N.S., in 1862. Two years later a mine was opened at Markhamville, N.B., which has proved the most productive of the district. The ore occurs as lenticular layers interbedded in limestone, or in pockets bearing from a few pounds to four thousand tons. Other localities are. Quaco Head, Jordan Mountain, Glebe and Shepody Mountain, N.B., and Cheverie, Walton, Onslow, Loch Lomond, Cape Breton, N.S. Much of the ore is sufficiently pure to be used for chemical purposes, some of it selling at the mines for $125 a ton. The lower grades are used in the iron industry. In Colchester and Pictou counties many of the iron ores are highly manganiferous. A number of deposits of wad occur in Quebec, principally in the Eastern Townships. Pyrolusite is found on the Magdalen Islands, Que., and manganite on the north shore of Lake Superior. New Brunswick and Nova Scotia are the only producing provinces, and most of their output is exported. In 1895 the production was 125 tons, valued at $8,464. In the same year oxid of manganese to the value of $2,800 was imported. The industry has fallen off enormously since 1890.

LITERATURE.—Penrose in the annual report, 1890, Vol. I., Arkansas Geol. Sur., gives a complete account of the origin, occurrence, use, etc., of the manganese deposits of America. Geol. Sur. Can., V. 1890 S. Dawson, "Acad. Geol."

CHROMIUM.

Chromium occurs in nature as the mineral chromite ($FeCr_2O_4$), isomorphous with magnetite. It is usually massive, finely granular or compact, hard and black. It occurs in serpentine, either in veins or in imbedded masses. It is rarely reduced to the metallic state, but a small quantity is used in a steel alloy, valuable on account of its great hardness combined with toughness. A more extensive use is in the manufacture of chromates of sodium and potassium used in dyeing.

Chromite occurs in Quebec in the neighborhood of the asbestos mines. Many pockets have been discovered and quarried, but no systematic mining operations have been undertaken. Much of the ore averages 50 per cent. and is worth at the railway $26 a ton. The richer ore is shipped to the United States and a small amount to Nova Scotia. The lower grade ores are marketed in Great Britain. The production in 1895 was 3,177 tons, valued at $41,000.

LITERATURE.—Geol. Sur. IV. 1888 K. Can. Mining Manual, 1896.

CHAPTER IV.

NICKEL AND COBALT.

Ores of Nickel.—There are a large number of minerals containing nickel, but most of them are not found in any abundance. Those which have been used as ores are a few of the sulfids and a silicate. Millerite (Ni S) contains 64 per cent. of nickel, and is characterized by its brass-yellow color, greenish-black streak and hair-like crystals. Niccolite is the arsenid of nickel and gersdorffite the sulpharsenid. Pyrrhotite, (Fe_7S_8) is, however, the chief sulfur ore of nickel. In many localities a small percentage (up to 6) of nickel replaces a portion of the iron. The nickel is, indeed, an impurity in the pyrrhotite, and only the large amount in which this mineral is found makes it valuable as a source of nickel. In color pyrrhotite is bronze-yellow to copper-red, and often tarnished on the surface. The streak is dark greyish-black, and the powder magnetic. Genthite is a hydrous, nickel, magnesium silicate found on Michipicoten Island, Lake Superior, and containing 23 per cent. of nickel. Closely related to it is garnierite, a soft, amorphous, pale-green mineral somewhat indefinite in composition but containing eight to thirty-six per cent. of nickel.

Distribution.—The minerals containing nickel are found all over the world, but in few localities are they sufficiently concentrated to be of value as ores. Pyrrhotite is found from the Atlantic to the Pacific, but the amount of nickel contained is usually small. Pyrrhotites from near St. Stephen, N.B., show 2.5 per cent. of nickel, which is almost as much as the average of the famous Sudbury region.

In the last-named district several score of rich deposits of nickeliferous pyrrhotite have been found in a belt of country four or five miles wide and fifty-five miles long. Outlying deposits occur south to the Georgian Bay, to the north-west at Straight Lake, and probably far to the north-east. Deposits of a similar character are worked in Norway. Millerite was noticed by officers of the Geological Survey at the Wallace Mine on the shore of the Georgian Bay as far back as 1848, but it was not until 1883 that the riches of the district were discovered.

Silicates of nickel are seldom absent from the magnesium rocks of the Eastern Townships, Que., but in no place are they of economic importance. They are reported in paying quantities from Oregon and Nevada, and small amounts have been mined. New Caledonia, a French penal colony, has until recent years been much the largest producer of nickel. The ore, garnierite, is found in veins in serpentine associated with chromic iron and steatite.

Millerite has been worked at the Lancaster Gap Mine, Pa., for a number of years, but the mine is no longer productive. The same mineral was mined at Brompton Lake, Que., but as the rock mass only contained 1 per cent. of nickel the operation was not profitable.

Geological Occurrence.—All the important deposits of nickel occur in metamorphic rocks. Garnierite, the silicate, is found with serpentine, and the sulfids and arsenids are associated with quartzites, slates and schists. In the Sudbury District the ores are found in masses, not in true fissure veins, in Huronian strata. The ore mass is usually a brecciated mixture of country rock, chalcopyrite and pyrrhotite. Sometimes one, sometimes the other of the last two predominates, but they are too intimately mixed to admit of separation by sorting. Originally the deposit was worked for copper. The ore mass is usually lens-shaped, not only horizontally but also vertically. Diabase and diorite have been erupted through the Huronian sediments, and the nickel and copper deposits are usually close to the contact. Occasionally the ores are found in granite where diabase has pierced it. The sulfids are often found in the diabase itself, and the enclosing rock is frequently impregnated. This leads to the conclusion that the ores and the diabase have been introduced at the same time, possibly at the close of the Huronian.

Several companies are energetically engaged in mining and roasting the ores of the Sudbury district. As mined the ore contains one to four per cent of nickel and four to ten per cent. of copper. About

3 per cent. of nickel seems to be the average, and occasionally one-fiftieth of this is cobalt. After being raised the ore is piled in heaps and roasted, sulfur being given off. It is then smelted to a matte which carries about 20 per cent. of nickel and 20 per cent. of copper. This is shipped to New Jersey or Wales for further treatment, as no refining is done in Canada.

Uses.—The metal is used for subsidiary coinage by the United States, Belgium and Germany. A small amount is made into cheap jewelry, principally watch cases. An alloy of nickel, copper and zinc is largely used under the name of German silver. Electroplating with nickel is widely used to beautify parts of stoves, bicycles, etc. A far more extensive use than any of these has been found in recent years. Steel, alloyed with a small percentage of nickel, is greatly increased in strength. For armor plate the alloy seems particularly adapted. Where lightness as well as strength is a consideration, nickel-steel seems destined to replace ordinary steel.

The price of nickel is gradually lessening as improved processes of refining are invented. In 1873 it was worth $6.00 a pound; in 1890, 65 cents; in 1893, 52 cents; in 1895, 35 cents. The annual consumption is about four thousand tons, of which Canada furnishes one-half; Norway mines a few hundred tons, and nearly all the remainder comes from New Caledonia. The Canadian production has been as follows:

YEAR.	Pounds of Nickel in Matte.	Value at Mine.	Final Value.
1893	3,983,000	$630,000	$2,071,000
1894	4,907,000	559,000	1,871,000
1895	3,889,000	522,000	1,361,000

LITERATURE.—Description of the Sudbury Deposits: Bell, "Report F, Geol. Sur. Can.," V., 1890-91; Barlow, "Rep. S, Geol. Sur. Can.," V., 1890-91, pp. 122-140. Metallurgy: Bureau of Mines of Ontario Rep. 1892, pp. 149-161. Use in Armor Plate, etc.: *Ib.* 1893 and 1894. Origin: "Mineral Industry," 1895, p. 746.

COBALT.

Cobalt occurs in a number of minerals, principally sulfids and arsenids, and usually associated with nickel and iron. Nearly all meteoric iron contains a small amount of the metal. While there are a number of minerals, they are not widely distributed and seldom occur in large amount. Most of the cobalt of commerce is a by-product in the refining of nickel. In one mine of the Sudbury district about 0.08 of 1 per cent. of the ore smelted is metallic cobalt. This represents a production of nineteen tons in 1893, and three tons in 1894, worth about $460 a ton. Cobalt is used, chiefly as the oxid, in the manufacture of paints, colored porcelain, etc.

CHAPTER V.

COPPER AND SULFUR.[*]

Ores of Copper.—*Chalcopyrite* (Cu Fe S_2), the most common ore of copper, resembles ordinary iron pyrites, but is much softer and of a deeper yellow. It yields, when pure, a little over 34 per cent. of copper. This is the chief copper ore of the Sudbury District. *Bornite*, also known as variegated copper ore, is an iron copper sulfid like chalcopyrite, but with a percentage of copper which varies from 55 to 60. It is copper-red to brown in color, and the surface is always tarnished. *Chalcocite* (Cu_2 S), called also vitreous copper ore, contains about 80 per cent. of copper. It is blackish lead-grey in color, often tarnished blue or green, and is comparatively soft. It is found in rich, but small, deposits in the Carboniferous rocks of Pictou, N.S. These three ores are said to furnish three-fourths of the world's supply of copper. *Native Copper* is next in importance, furnishing about one-sixth. Most of this comes from the south shore of Lake Superior, but the mineral is also found in considerable quantities on the north side. It is found also in the Triassic trap of Nova

---

[*] " Standard " Dictionary.

Scotia, on the Coppermine River far to the north, and in British Columbia, but so far not in economic amounts. The mineral has a characteristic red color, a bright metallic lustre, and can be cut with a knife. *Malachite*, the green carbonate of copper, *Azurite*, the blue carbonate, *Cuprite*, the red oxid, *Chrysocolla*, the bluish-green silicate, are other ores as yet of no economic importance in Canada. *Tetrahedrite*, also called grey copper, is a complex sulfid of copper, antimony and other metals. It is proving of value as a source of silver in British Columbia, and so incidentally yields copper.

Geological Occurrence.—Copper ores are more usually found in the oldest rocks, the Archæan and Cambrian strata being particularly rich. Workable deposits are, however, found as late as the Permian, as at Mansfeld, Germany.

The ores are found (1) in veins intersecting older rocks, as at Bruce Mines, north of Lake Huron; (2) in mass deposits, as at the immense quarries on the Rio Tinto, Spain; (3) disseminated in beds, as at Mansfeld; (4) as impregnations in amygdaloids and conglomerates, well exemplified in the basin of Lake Superior.

Canadian Localities.—*Maritime Provinces.*—The copper ore mined in Canada at present is only incidental to the production of sulfur, nickel and the precious metals. At a number of places in the Maritime Provinces development work has been undertaken. Sulfids have been found in Pictou county, N.S., and in St. John and Albert counties, N.B., and in the latter case were worked for a time. In

Annapolis county the Triassic traps contain strings of native copper which may prove of value. The Coxheath Mine, Cape Breton, is of greater promise. A number of veins bearing chalcopyrite are there found traversing a mass of felsitic rocks of Laurentian age. Considerable sinking and drifting has been done, and several thousand tons of ore have been raised, large parts of which average 10 per cent. of copper. Smelting works are being erected on Sydney Harbor. About two thousand tons of copper ore are mined annually in Newfoundland.

Quebec.—Several score of "mines" and many more "prospects" have been partially explored in southeastern Quebec. Some of these have proved to be rich deposits, and others might probably have been made paying investments had development work been carried far enough. The deposits occur along three anticlinal axes running north-eastward from the Vermont boundary. The ores are the sulfids—chalcopyrite, chalcocite and bornite. They are found in veins, in irregular masses and in what seem to be beds, but which are probably in reality of eruptive origin. In nearly all cases they are associated with diorites, apparently of Cambrian age. In the western belt the variegated and vitreous ores are most common, and occur in dolomitic beds belonging to the Upper Cambrian. The pioneer mine of the district was the Acton, first worked in 1858. From it sixteen thousand tons of 12 per cent. copper were taken.

In the central and eastern belts the ores occur in Pre-Cambrian, micaceous and chloritic schists. The

Harvey Hill and Huntingdon mines represent the former region, the Capleton group the latter. Many hundred tons were produced by the Harvey Hill and Huntingdon, but they have been closed for several years. In the Capleton district the ore is a mixture of chalcopyrite and pyrite containing thirty-five to forty per cent. of sulfur, and four to five per cent. of copper. It carries in addition from one to seventy-five ounces of silver to the ton, averaging $4.00 to $5.00 in value. The Eustis mine, typical of the group, is an irregular deposit four to fifty feet wide and explored to a depth of 1,600 feet. Most of the ore is shipped to New Jersey for the manufacture of sulfuric acid. The copper and silver are afterwards refined.

Ontario.—Chalcopyrite and native copper are the two important copper ores of Ontario. The former occurs in greatest abundance north of Lake Huron; the latter around the shores of Lake Superior. The years 1849-1875 constitute the first period of copper mining in Ontario, during which much ore was raised and shipped, but without profit to the shareholders.

The Bruce and Wellington mines on the north shore of Lake Huron produced nearly forty-five thousand tons of dressed copper ore, worth about $3,500,000. The mines embrace half a dozen veins of quartz in diorite, spread over an area of a square mile. The veins were three to fifteen feet wide, and the workings were carried down about 450 feet. The ore, mainly chalcopyrite, averaged $6\frac{1}{2}$ per cent. copper as it came from the shaft. The great expense of mining

THE MINERAL WEALTH OF CANADA. 59

and shipping to England, the failure of smelting plants erected at the mines, the decrease in the value of copper, all contributed to make the work unprofitable at that time.

Since 1846 a number of companies have made explorations at Michipicoten Island, St. Ignace Island, Mamainse, Point Aux Mines, and other places on the north shore of Lake Superior. The rocks outcropping at these points are the same as those which in Michigan have proved to be so rich in native copper. According to Irving, the bed of Lake Superior is a geosyncline, the Huronian and overlying Keweenawian rocks extending beneath the waters of the lake in a gentle fold. The Keweenawian formation, or Nipigon, as it is known in Ontario, outcrops as a narrow fringe around part of the shore of the lake, except in the vicinity of Lake Nipigon where a considerable area is found. Through these Nipigon sediments immense masses of volcanic material were erupted, and in the more vesicular outflows and in the associated sandstones native copper is now found. Keweenaw Point on the south shore has proved to be exceptionally rich. One of its mines, the famous Calumet and Hecla, produced, in 1895, one tenth of the whole world product of copper. On the Canadian shore native copper has been found at a number of points, often in rich though small amounts, and always inciting the explorers to develop their properties further. The ore exists as an impregnation of beds of sandstones, conglomerates and vesicular trap. It is also found in veins, associated with calcite, cutting these beds. The copper is always irregularly distributed, and considerable quantities of barren rock have often to be removed. Prehnite and epidote are here associated with the copper, as on the south shore. Indeed, the indications are quite favorable, but so far no profitable mine has been discovered. A six-hundred-pound mass of native copper, taken from a shaft at Mamainse, is probably the largest yet found. At Michipicoten a shaft has been sunk over five hundred feet in an amygdaloidal bed, and 1,500 feet of drifting done. The copper carries a little native silver in many places, and malachite, cuprite, chalcopyrite are often found with it.

In 1882 large deposits of chalcopyrite were discovered near Sudbury, Ont. The ore is "a brecciated or agglomerated mixture of the pyrrhotite and chalcopyrite along with the country rock." This mixed ore is usually in or near masses of diorite, intrusive through Huronian or Laurentian rocks. It occurs in lenses which thicken and thin out vertically as well as laterally. At first the ore was mined for copper, but nickel, which is found in the pyrrhotite, is now the more valuable constituent. (See Nickel.) The average output of the three mines of the Canadian Copper Company is 4.3 per cent. of copper, and 3.5 per cent. of nickel. The ores are roasted in heaps in the open air to drive off sulfur, then smelted to a matte containing eighteen to twenty per cent. each of copper and nickel. This matte is shipped to New Jersey or to Wales for further treatment. The quantity of ore in the district seems inexhaustible, and the copper and nickel mines are now firmly established. In 1895 eighty-six thousand tons of ore were smelted at Sudbury.

British Columbia.—Ores of copper are widely distributed throughout the whole area of the Pacific Province. Several attempts have been made to develop them, but so far unsuccessfully. Many of the most promising gold and silver ores contain large amounts of copper, and the recently developed mines in West Kootenay are yielding a very large amount of copper in addition to the gold and silver for which they are worked.

Uses.—Next to silver, copper is the best conductor of electricity, and so is used in telephone trunk lines, trolley wires, etc. Its great toughness makes it valuable for boilers, stills, sheeting wooden ships, etc. It is a component of brass, bronze and other alloys used for machinery, cannon, bells, coins and statuary. A number of its salts, as blue vitriol and Paris green, find extensive use in the arts.

Production.—No copper is at present refined in Canada, all the ore mined being exported either as raw ore carrying about 4 per cent. of copper or as a matte carrying fifteen or twenty per cent. In 1894 the final value of the copper in the ore produced was $736,000, of which Quebec contributed $207,000; Ontario, $495,000; British Columbia, $34,000. In 1895 the total value was $949,000, the increase being due to the copper-gold mines of British Columbia. In 1896 the output of this province was doubled, and the total for the year is a little over a million. The imports of pig and scrap copper in 1895 were valued at $7,000, and of manufactures at $252,000. The annual production of copper in the world is steadily increasing, the increase being just about equal to that made by the United States. The following table is compiled from Rothwell's "Mineral Industry":

PRODUCTION OF COPPER, 1895.

	Metric Tons, of 2,204 lbs.
Australasia	10,000
Canada	3,987
Cape of Good Hope	30,000
Germany	17,000
Japan	19,000
Mexico	12,000
Russia	5,000
Spain and Portugal	56,000
United States	175,000
All others	12,000
Total	340,000

LITERATURE.—Localities and History of Operations.—Maritime Provinces: Dawson's "Acadian Geology." Quebec: Geol. Sur. Reports, 1863; III. 1887 K; IV. 1888-89 K; Obalski, "Mines and Minerals of Que.," 1890. Ontario: Lake Superior —"Geol. Can.," 1863; Geol. Sur. III. 1887, 9-12 H; "Min. Resources of Ont.," 1890; Rep. Bur. of Mines, Ont., 1893; Bruce Mines, etc.—"Min. Res. Ont.," 1890; Sudbury—Geol. Sur. Rep., V. 1890 F. (See also references under Nickel.) British Columbia: Rep. Geol. Sur. III., 1887, 101 R, 152 R; VII. 1894, 52 S. Production.—Reports of the Geol. Sur. of each year. Irving, "The Copper-bearing Rocks of Lake Superior." Peters, "Modern American Methods of Copper Smelting."

SULFUR.

Sulfur, from a chemical standpoint, is an acidic element, and so in strictness should not be classed here under the metals. As, however, it is mined in Canada as a constituent of copper ores, this is a convenient place for considering it. Sulfur is found native at only a few places in Canada, and never in economic quantities. It does exist, however, in immense quantities as sulfids of a number of metals. Pyrite ($Fe\ S_2$), the sulfid of iron, contains 53 per cent. of sulfur. It is a brassy-looking mineral, hard enough to strike fire with a piece of steel, and is frequently found in cubic crystals. It occurs in rocks of all ages, and as it oxidizes readily it frequently causes undesirable stains on building stones. Chalcopyrite ($Cu\ Fe\ S_2$) is a similar mineral, but softer and yellower. It contains 35 per cent. each of copper and sulfur. These two minerals are largely used as sources of sulfur for sulfuric acid. Other sulfids occurring in large quantities in Canada are galena (PbS), the sulfid of lead; blende (ZnS), the sulfid of zinc; pyrrhotite ($Fe_7 S_8$), another sulfid of iron.

Uses.—Sulfur is required for manufacturing gunpowder, matches and vulcanized rubber; for bleaching straw and woollen goods; for cementing iron and stone; for making sulfuric acid. This last is one of the most important compounds known to chemistry and commerce. It is said that a nation's civilization may be gauged by the amount of sulfuric acid it consumes.

Although native sulfur is required for most purposes, pyrite answers equally as well as the element in making sulfuric acid. The pyrites, iron and copper, are consequently slowly driving the native element from the acid factories by reason of their cheapness. Especially is this true of ores like those of Capleton, Quebec, which are valuable for their copper and silver contents, and from which the sulfur must be separated anyway.

The pyrites are burned to form sulfur dioxid gas, and the residues are treated with acids to obtain the copper, silver or gold. Thoroughly burned pyrite retains about 1 per cent. of sulfur, and iron containing not more than that can now be used for some purposes. Pyrites suitable for sulfuric acid should have the following characteristics: (1) A high per cent. of sulfur, 35 to 53; (2) freedom from arsenic, antimony and lead; (3) readiness in yielding the sulfur; a granular and porous pyrite is easier to work than a compact one; absence of fluxes is desirable; (4) valuable accessory metals, as silver, copper, gold, are a great advantage.

Production.—The Capleton and Eustis mines in southern Quebec are the only Canadian producers which use the sulfur in their ores. A part is made into sulfuric acid at the works; a much larger portion is shipped to the United States. A third portion is smelted at the mines, the sulfur being wasted and the matte exported. These mines are described under Copper, earlier in this chapter. Other sulfuric acid factories at Brockville and at Smith's Falls,

Ontario, have also used pyrites. Immense quantities of sulfur are wasted at Sudbury. Nearly five million pounds of sulfuric acid are used annually in refining Canadian petroleum.

		1890.	1895.
Production of	Tons.....	49,000	34,000
Pyrites....	Value....	$123,000	$103,000
Imports crude	Tons.....	2,220	2,450
Sulfur.....	Value....	$44,000	$57,000

LITERATURE. — "Min. Resources of Ont.," 1890. Rep. Geol. Sur., 1874, p. 304; *ib.* IV., 1888, 53 K, 158 K; *ib.* VIII., 1895, S.

CHAPTER VI.

GOLD AND PLATINUM.

GOLD.

IN the first half of the present century Russia held first place as a gold producer. In 1848 came the discoveries in California, which soon sent the United States to the top. Three years later rich deposits were announced in Victoria. In a few years Australia climbed to the foremost position, and the place of honor has alternated between that island and the United States until recently. The South African field, discovered in 1884, has been developed with surprising rapidity. In 1895 the Transvaal succeeded in passing Australia, and if the rate of advance is continued it will soon surpass the United States.

In 1896 announcements were made of rich discoveries, which it is hoped will make Canada a worthy rival of California, Victoria and the Transvaal. In 1895 Canada was twelfth among nations in the value of her gold output, and it is quite probable that she may reach fifth, or sixth, place within a few years. Mexico, which at present ranks fifth, is increasing her gold output very rapidly. On the same Cordilleran range as British Columbia, with enormous deposits of silver already exploited, Mexico may

THE MINERAL WEALTH OF CANADA. 67

prove as rich in gold as Canada. It will be some years before either country reaches the fourth place now held by Russia.

Origin.—All substances can be resolved into one or more of the seventy primary elements. These elements, of which gold is one, cannot be changed into one another, though they combine in various proportions to form different substances. So far as we know they have existed from the creation. On the cooling of the molten earth most of them assumed a solid condition, either alone or in combination. Gold seems to have remained free, and pretty thoroughly distributed through the crystalline rocks. It is found now in nearly all rocks, and in sea water, but in such minute quantities that it cannot be economically recovered.

Nature, however, at once set about concentrating it for man's use when he should appear in later ages. Running water was the agent employed. The ancient rocks were slowly disintegrated and the minerals floated off. Gold, which is seven times heavier than quartz, was carried down the turbulent mountain streams, to be deposited with the coarser sands and gravels at the first eddy or level stretch of water, whilst the lighter minerals and finest particles were carried on. Many of these river sediments, perhaps reassorted by lake or ocean action, have been consolidated by pressure to form sandstones and conglomerates. Finer particles of gold were even carried to the sea, so that marine sediments also contain disseminated gold, though in exceedingly minute amounts. Subjected to pressure and heat these sediments became the metamorphic rocks—slates and schists.

Meanwhile, in another way, concentration was being effected. Fissures were made in the metamorphic, and also in the igneous rocks. Hot solutions of quartz, carrying iron and copper sulfids, leached the gold from the underlying and adjacent rocks and placed it in the vein where the quartz and pyrites solidified around it. These quartz veins have also been subjected to denuding agencies, and they probably have furnished most of the gold found in modern river gravels. In still a third way has concentration been brought about. In many copper and silver mines gold is an accessory mineral. These deposits are sometimes of an eruptive origin, *i.e.*, the mineral matter has come from below in a fluid condition.

Occurrence.—Gold nearly always occurs as the native element; its natural compounds are mineral curiosities. It alloys readily with silver, and is nearly always found with a small percentage of that metal. Quebec gold contains about 12 per cent. of silver; that of Nova Scotia is nearly pure. When in visible particles, gold is easily recognized by its yellow color, malleability, and by the ease with which it may be cut with a knife. Iron and copper pyrites, the former known as "fool's gold," are the only minerals which resemble it. Both are much harder, both crumble under a hammer, both yield fumes of sulfur when heated with a blowpipe, and both lack the peculiar lustre of gold.

Dependent on the mode of origin, four classes of gold deposits may be noticed:

1. Placers, in which auriferous gravels of the Tertiary and Quaternary ages are worked. The gold is free, and may be separated easily from the sand by means of mercury. These placers have been, and probably still are, the most important source of gold. Their place is, however, slowly being taken by the next class.

2. The second class of deposits are the auriferous quartz veins. They are widely distributed in all kinds of metamorphic rocks of all geological ages. They are more expensive to work than the first, since the ore must be mined and crushed before being amalgamated. Two subdivisions should be noted: (*a*) That in which the gold is free in a quartz containing little or no sulfids; (*b*) That in which a considerable part of the gold is in sulfids of iron, copper, lead or zinc in the quartz. This class, especially division *a*, is represented by the ores of Nova Scotia and western Ontario.

3. The ancient gravel deposits, as illustrated by the auriferous sandstone of Cambrian age, in the Black Hills, Dakota. The Carboniferous conglomerates of Australia, and also of Nova Scotia, are other examples.

4. The occurrence of gold in eruptive deposits makes a fourth class. The ores of the Rossland (B.C.) region are an example.

Methods of Milling.—The methods of separating a metal from its ore hardly find a place in a work of this kind. A brief explanation may, however, be given for gold, and details can be sought in a work on metallurgy. Free gold is easily separated from its gangue. In placer mining an inclined trough is arranged near the supply of gravel. Across the bottom are placed cleats, and over them a stream of sand, water and gold is caused to flow. These cross-pieces in the bottom of the sluice check the current, and so tend to hold the heavy gold which is sliding along the bottom. Behind these cleats or riffles mercury is placed. This element has a great affinity for gold, and greedily grasps and dissolves any particle being washed over it. At intervals the amalgam of mercury and gold is placed in a retort, and heated to drive off the mercury. The gold, left behind as a powder, is fused and sent to the market as a "brick."

In quartz mining the first step is the crushing of the ore in a stamp mill. Iron weights or stamps of eight hundred pounds are dropped about eight inches, about eighty times a minute, on pieces of quartz. Water carries off through a sieve the fine pulp, which then flows over an inclined copper table covered with mercury. At intervals the amalgam is scraped off and retorted as previously described.

Any gold held in the sulfids is not attacked by the mercury, and so passes over into the tailings and is lost. To prevent this, a mechanical separation of the heavy sulfids and light quartz is effected. A machine known as a vanner is largely used. A wide belt constantly moves upward over an inclined table.

THE MINERAL WEALTH OF CANADA. 71

The stream of pulp is directed on this belt which carries up the heavy sulfids containing the gold, while the water carries downwards the light quartz. In this way the "concentrates" are saved for further treatment. These concentrates, and in many mines the whole ore, must be treated chemically to obtain the gold. First they are roasted, or calcined, to free them from sulfur. Then they are treated with chlorin, potassium cyanid, or bromo-cyanogen to dissolve the gold, which is afterwards precipitated.

Canadian Localities.—*Nova Scotia*—Along the Atlantic coast of Nova Scotia there is an extensive development of Cambrian strata. The rocks, which are quartzites, sandstones and slates, are about twelve thousand feet thick, of which the lower three-fourths are most auriferous. At many localities igneous rocks have been erupted, and apparently at the same time quartz veins were formed. The sedimentary strata were thrown into folds with their axes running east and west. Along the denuded crests of these folds quartz veins are found which resemble bedded deposits. These veins are for the most part narrow, most of those worked being less than a foot in width. They extend from a few hundred feet to several miles in length. The area through which they are found is probably six thousand to seven thousand square miles, though actual operations are restricted to a much smaller area. The ore is almost entirely free milling, and has averaged $13.70 a ton for the province for thirty years. The Gold River and Renfrew districts have the richest ores at present.

The Stormont and Caribou districts, working on low grade ores, yield the largest returns. The total production to the end of 1894 was $11,000,000. The Sherbrooke and Waverly districts have been the chief producers. Nova Scotia seems destined to yield a small but steady supply of gold. The industry is being extended to the low grade ores which exist in much wider veins, and which are being mined and milled for $2.50 a ton, leaving all over that for profit. This is small in comparison with Rossland, B.C., where $15 ore is the least that will pay at present.

Quebec.—Gold was accidentally discovered in the Gilbert Creek, a tributary of the Chaudière, about 1823. For many years it was neglected, and the mining operations even of the last fifty years have been very desultory. The gold is found in gravels which constitute the beds of preglacial streams. The Gilbert, Des Plantes, DuMoulin, DuLoup tributaries of the Chaudière in Beauce county have been the chief producers. Ditton Creek has also proved to be rich. The gravels which lie on bed rock are always richer than those above a bed of clay. Many of these early gravels are one hundred feet below the level of the present streams. They are covered by boulder clay, a product of the glacial age. The gravels are always richer when near veins of quartz which intersect the Cambrian rocks of the district. These rocks, which are slates and sandstones, closely resemble the corresponding gold-bearing strata of the same age which occur in Nova Scotia. Workable quartz veins have not yet been discovered. The gold is all derived from the placers, much of it in a primitive way. Modern hydraulic methods are being applied, and the output, which has been small and uncertain, will doubtless be increased.

Ontario.—Rich deposits of free gold were discovered in Hastings county in 1866. Prospectors flocked in and located hundreds of properties. Many companies were formed and development work begun. The first returns were very encouraging, but at a slight depth the ore changed from a free-milling quartz to a refractory arsenical pyrites. With the methods in use the gold passed over into the tailings and was lost. No successful means of separating the gold could be found, and one after another the mines were closed. Within the last few years renewed attempts have been made with more modern processes, with the probability of final success. Besides these rich arsenical ores free-milling quartz veins have been worked not only in Hastings, but in Peterboro' and in Addington. The Hastings district will likely become a small but steady producer. The veins occur in Upper Laurentian or Huronian strata.

In the strip of Huronian rocks stretching northeastward from Lake Huron to Sudbury, and on to the Ottawa River, a number of promising gold discoveries have been made. For the most part the ore is a free-milling quartz with a little pyrite, occurring in bedded deposits. Two stamp mills have been erected, but the output is irregular as yet.

From Lake Superior west to Manitoba prospecting has been carried on vigorously since the opening of the railway. Many hundred "prospects" have been located and considerable development done. About a dozen mines are equipped with stamp mills, several of which have passed the experimental stage and are working continuously. The ore is a free-milling quartz, containing about 2 per cent. of sulfids. In mill tests the ores give from six to thirty dollars in free gold, and about one-fifth more in the concentrates. The veins, which are both bedded and fissure, occur usually in Keewatin (Huronian?) schists, but also in the Laurentian granite near the contact of the two. The Sultana, the best developed mine in the district, is right at the contact of the Keewatin and Laurentian. The shaft is over 350 feet in depth, and the vein has a width, on the third level, of upwards of 30 feet.

British Columbia. — After gleaning the surface riches of California the gold hunters drifted northwards. In 1857 came the first authentic news of rich finds on the Fraser. The next spring 20,000 people reached Victoria within four months. The difficulties of penetrating the interior were, however, so great that the majority turned back. A few thousand people pushed up the Fraser and were richly rewarded. Their methods were crude in the extreme, and only the richest bars proved profitable. Year by year they pushed farther up the main stream and its tributaries, carrying with them all the necessaries of life. In 1860 they reached the Cariboo district, one of the best placer mining camps ever found. The following year came the discovery of Williams and Lightning creeks, on which were found the richest placers yet discovered in British Columbia. In two years it is said $2,000,000 were got out by 1,500 men. The richness of "Golden Cariboo" caused a large immigration from all parts of the world for the next few years. A party started overland from eastern Canada, and after many misfortunes most of them reached their destination. Placers were next found on the Kootenay and Columbia. Northward from Cariboo the prospector forced his way into the Omenica district. A few years later the advance guard reached the Cassiar district on the northern boundary of the province. In 1880 the tide, in a restricted flow, had reached the head-waters of the Yukon in the North-West Territories.

These early pioneers skimmed but the surface of the bars, *i.e.*, the portions of the river bed uncovered at low water, and of the terraces on the banks. Succeeding miners sought the equally rich deposits more difficult of access. For instance, Lightning Creek was filled with glacial deposits to a depth of 50 to 150 feet. As the modern stream bed was rich there was a probability of the preglacial bed being the same. Shafts were sunk and tunnels run, and the old channel cleared out for a distance of three miles. Again, auriferous gravels on the banks of streams are now mined by hydraulic methods where the old "rocker" would not pay. Streams of water under great pressure are directed against a gravel bed, and everything washed down a sluiceway at very small cost. Riffles in the sluice catch the gold.

About $58,000,000 of gold have so far been taken

from the placers of British Columbia, principally from river bars. Many millions are yet to be obtained by hydraulic methods from the terrace deposits of the Fraser and other streams. Even the beds of the rivers may be successfully exploited by dredges or by dams. There is scarcely a stream of importance in British Columbia in which "colors" of gold cannot be found. The richest areas are, however, in the parallel and partly overlapping ranges collectively known as the Gold Range. This range includes the Purcell, Selkirk, Columbia or Gold, Cariboo, Omenica and Cassiar Mountains. It lies parallel with the Rocky Mountain range to the south-west. It will probably prove the most important metalliferous belt of British Columbia. The Vancouver range is also very promising.

Since 1863, when they yielded nearly $4,000,000, the placers have been steadily decreasing in value. In 1893 the product was only $380,000; but in 1896 it rose to $544,000. Attention was for this reason directed to the quartz veins from which the gold was derived. The province is passing through the experience of California and Australia, where the miners began on placers but are now working the veins. The most important mines and "prospects" are in the Trail Creek division of the West Kootenay district and in the Kettle River and Osooyos divisions of the Yale district. In some camps of the last named the ore is a free-milling quartz. In the Trail Creek division nearly all the ore is refractory. The three divisions lie side by side along the northern boundary of Washington state, and are much alike in the character of their ore. Greater development has taken place in Trail Creek owing to accessibility. The typical ore of Rossland is either a " nearly massive fine-grained pyrrhotite and copper pyrites, with more or less quartz and calcite," or a poorer ore consisting of diorite with a comparatively small percentage of the sulfids. The ore resembles that which carries nickel at Sudbury. Average smelter returns of the Le Roi mine are for first class ore $53; for second, $28. This includes the silver and copper values. Immense bodies of low grade ores are found, and the successful treatment of these will depend on a cheapening of the smelting process. The production of the division has increased with enormous rapidity, and everything points to Kootenay as an enduring and profitable mining district. Quartz veins will in time be worked in Cariboo to the north, but at present that district depends on its placers. The district of Alberni on Vancouver Island may also prove rich in quartz deposits. With the advent of more modern methods of working the placers, and with the new developments in vein mining the output of gold from British Columbia is bound to increase enormously.

Other Placers.—Far to the north, on the Yukon and its tributaries, miners are washing the sands by the old methods and are reaping an enormous harvest. On the Saskatchewan also, near Edmonton, similar work is in progress. It is of interest to note that here the immediate source of the gold is the Cretaceous sediments of the Edmonton series. These sandstones were probably derived from the Coast Range of British Columbia.

The following tables are self-explanatory:

GOLD PRODUCTION OF CANADA.

PROVINCE.	1893.	1894.	1895.	1896.
Nova Scotia.......	$381,095	$377,169	$ 406,765	
Quebec	15,696	29,196	1,282	$1,022,000
Ontario......	14,637	39,624	62,320	
Alberta and Yukon.	185,640	140,000	150,002	
British Columbia...	379,535	456,066	1,290,531	1,788,206
Total.........	$976,603	$1,042,055	$1,910,900	$2,810,206

GOLD PRODUCTION OF THE WORLD, 1895.

1. United States.................. $46,800,000
2. Transvaal..................... 43,000,000
3. Australasia 42,800,000
4. Russia 34,000,000
5. Mexico....................... 5,600,000
6. China....................... 4,700,000
7. India 4,500,000
8. Colombia.................... 3,200,000
9. Germany.................... 2,900,000
10. Brazil 2,200,000
11. British Guiana 2,200,000
12. Canada..................... 1,900,000
13. Austria 1,800,000
14. French Guiana 1,600,000
    All others 6,243,772

$203,443,772

—*Rothwell's " Mineral Industry."*

LITERATURE.—General: Phillips and Louis "Ore Deposits;" Rep. S of the Geol. Sur. for each year; Can. Mining Manual, 1896. Nova Scotia: Annual reports of Dep. of Mines; Trans. Am. Inst. Min. Eng. XIV.; Trans. Min. Soc., Nova Scotia, 1894-95; Bibliography 158 P Geol. Sur. II. 1886. Quebec: Rep. K. Geol. Sur. IV. 1888. Ontario: Geol. Sur. 1871 ; Min. Resources of Ontario, 1890; Bur. Mines Rep. 1893, '94, '95,' 96. British Columbia: A brief history of the placer mining, a list of localities, references to literature are given in Rep. R Geol. Sur. 1887-88 ; ancient placer deposits and conditions of occurrence of recent ones, 310-329 B. Geol. Sur. VII. 1894. Bulletin No. 2 on Trail Creek in Rep. Min. of Mines, British Columbia, 1896.

PLATINUM.

Platinum is a silver-white metal, nearly always found alloyed with iron, rhodium, iridium and osmium. It is found as black grains in many gold placers. From magnetite it may be distinguished by its high specific gravity and its malleability. Like magnetite it is often magnetic. Its use in the arts depends on its great resistance to heat and to chemical reagents. Made into pans it is used in the concentration of sulfuric acid. In chemical laboratories it is used as crucibles, tongs, in galvanic batteries, etc. It is extensively used as a conductor of electricity in incandescent bulbs. It finds further employment in dentistry and in photography. Indeed, a substance so indestructible is only limited in use because of the high price.

Placer deposits of the Urals furnish most of the supply. Smaller amounts come from Colombia, Oregon and British Columbia. In 1891, the last-named furnished $10,000, but that amount has dwindled to

$3,800 in 1895. It was furnished by the streams of the Similikameen district. It is also found on the Fraser, Tranquille, Yukon, Saskatchewan and Chaudière, associated with the gold. As it does not alloy with mercury it is usually unnoticed. No doubt it could be found in many places in paying quantities. It seems to be connected in origin with masses of chromite and serpentine, and these again are associated with eruptive diorites.

CHAPTER VII.

SILVER, LEAD AND ZINC.

Ores of Silver.—Silver, lead and zinc are frequently associated in nature, and so they are best treated together. Silver is found native in many regions in small amounts. It is easily known by its pure white color, though it may have a dark tarnish. More commonly it is found in combination with other elements. With sulfur it forms argentite, blackish lead-grey in color, soft and malleable. With sulfur and antimony silver forms stephanite, iron-black in color, and pyrargyrite, dark red to black. Proustite is the corresponding arsenic compound, light red in color. All of these minerals are soluble in nitric acid, and yield a white precipitate on the addition of a solution of salt. Cerargyrite, or horn silver, is the naturally occurring chlorid. It is greyish-green in color, and looks like wax or horn. Silver always accompanies galena, the sulfid of lead, varying from a few thousandths of a per cent. to 1 per cent. With the larger amounts the mineral becomes an ore of silver. In a similiar way tetrahedrite, or grey copper ore, often carries enough silver to be of value as a source of that metal as well as of the copper. Still another source of silver is as an alloy with gold, most placer gold containing several per cent. of the white metal.

These silver minerals are rarely found in any amounts, but more commonly as strings and thin seams disseminated through a large bulk of gangue, mostly quartz or calcite. An ore mass yielding $100 to the ton is considered a rich deposit, and yet this is equal to only one-half of 1 per cent. of silver. It thus often happens that the silver minerals are in such small particles that they cannot be readily determined.

The greater part of the silver of the world is obtained as a by-product in the mining of other minerals. This is especially true of Europe and North America. Lead is the most common associated metal, though copper and zinc occur very frequently. In hundreds of mines operations would not pay were silver the only metal to be won.

Occurrence.—Silver ores occur in most of the classes of deposits tabulated in a preceding chapter. True fissure veins are perhaps most common, though bedded and contact veins are often found. These veins cut eruptive granite and Archæan schists in the Slocan district of British Columbia, and are found in sedimentary argillites of Lower Cambrian age in Ontario. Many of the most famous veins are incased in volcanic rocks of Tertiary age. The Comstock lode of Nevada which has yielded $325,000,000, occurs at the contact of two igneous rocks, and was evidently filled in very recent times by solutions from below. Associated with lead, silver occurs in mass deposits in

THE MINERAL WEALTH OF CANADA. 83

many parts of western America. The Cordilleras and the Andes, the backbones of the two Americas, are the great repositories of silver ores. Five-sevenths of the world's output comes from these regions.

Canadian Localities.—Argentiferous galena is reported from a number of places in Quebec, but apart from prospecting no mining operations have been undertaken. In Beauce and Compton counties quartz veins carrying galena are found cutting Lower Cambrian slates. This galena is frequently rich in silver. A large deposit of silver-lead occurs on the east side of Lake Temiscamingue. The copper ores of the Ascot belt usually carry silver up to ten ounces a ton. The average obtained from the Capleton pyrite mines is three to four ounces a ton, and this is the source of the present output of Quebec. (See Chapter V.)

In Ontario, at the west end of Lake Superior, there is a triangular area of Animikie rocks of Lower Cambrian age. These rocks are argillites and cherts, with intrusive sills of basic rocks which frequently appear as a capping of the precipitous hills. Associated with these trap-flows and their accompanying dikes are veins carrying silver ores. The gangue material is quartz, barite, calcite or fluorite, and blende, galena and pyrite are irregularly distributed in it. The silver occurs as argentite or as native silver, usually with the blende.

The most famous mine of the district is that of Silver Islet, discovered in 1868, from which some $3,250,000 were taken. The original islet, only 90 feet in diameter, lies near Thunder Cape in Lake

Superior. It owed its existence to a hard dike of quartz diabase over 200 feet in width, which resisted erosion. This was crossed by a vein striking N. W., which was thought to be traced to and on the mainland for about 9,000 feet. Where the vein crossed the island dike it was enormously rich, but in the argillites and where it crossed twenty other dikes, no paying ore could be found. The shaft was sunk 1,250 feet, and several bonanzas struck, with much barren rock between. The mine has been flooded since 1884, and it is doubtful whether it could be successfully operated again, though it is probable that similar ore bunches exist at greater depths. Many other properties in the district have been worked, but none approach this one in magnitude. The fall in the price of silver in 1892 caused the cessation of silver-mining in the Thunder Bay district. Argentiferous galena has been mined at Garden River, north of Lake Huron, and promising prospects are known in Hastings and Frontenac counties, and around Lakes Temagami and Temiscamingue.

British Columbia is the silver-producing province of the Dominion. From ten to twenty-five per cent. of the placer gold is said to be silver, but the value of it is accredited to gold in the tables published. Notwithstanding this the output of silver surpasses that of gold. Kootenay is the only producing district, the Ainsworth, Nelson and Slocan divisions being the chief regions, and smaller amounts coming from Trail Creek and East Kootenay. The Slocan is the most productive mining district in the province, its pre-

THE MINERAL WEALTH OF CANADA. 85

eminence being due to its silver-lead ore. Many of the veins are narrow, varying from two inches to twenty in width. Much of the ore is, however, very rich, and only this has made possible the opening and developing of properties so far removed from supplies in the face of a great fall in the value of silver. The average return on 18,000 tons of ore mined in the Slocan in 1896 was 117 ounces of silver and 53 per cent. of lead.

The chief ore is argentiferous galena, with some zinc blende and grey copper in a gangue of quartz and spathic iron. The veins cut across lower Palæozoic stratified rocks, and through eruptive dikes. They are also found in an extensive area of eruptive granite. Veins containing argentiferous tetrahedrite, or grey copper, are also found. Veins carrying argentite with native silver and gold in a quartz gangue are found in some granite areas. At the Hall's mines, in Nelson, the ore is a mixture of copper sulfids carrying silver. Smelters are at work at Nelson, Trail and Pilot Bay, but much of the ore is exported to the United States for treatment. The production is increasing rapidly, and with cheaper supplies many lower grade ores can be successfully worked. While only a small district is at present being developed, the argentiferous region extends 1,200 miles to the north.

Use and Production.—The use of silver is determined by its beauty, its comparative rarity, and by its resistance to the ordinary processes of change or destruction. Accordingly it finds employment in articles of luxury and ornament, and as a medium of exchange. For these purposes its hardness and durability are increased by the addition of seven and a half to twenty-five per cent. of copper. Owing to the greatly increased production of recent years and to the ease with which it may be won, the market price of crude silver has fallen greatly. The United States coining value is at the rate of $1.293 an ounce, while the average market value in 1895 was only 65 cents. An ounce of gold makes $20.67, so that at American coinage rates an ounce of gold is worth only sixteen of coin silver, but would purchase thirty-two on the market. In the following tables it is observable how the fall in the price of silver affected the production in Ontario. The annual production of the world is steadily increasing, although the total value is not so high as in 1890-93. Canada, which now ranks eleventh, will probably reach seventh place in the near future.

CANADIAN PRODUCTION.

—	1891.	1894.	1895.	1896.
Quebec	$182,000	$64,000	$53,000	$47,000
Ontario	221,000	....	....	....
British Columbia	3,000	470,000	1,105,000	2,101,000
Total	$406,000	$534,000	$1,158,000	$2,148,000
Value per oz...	0.98	0.63	0.65	0.67

SILVER PRODUCTION OF THE WORLD, 1895.

1.	Mexico	$33,225,000
2.	United States	30,254,000
3.	Bolivia	13,500,000
4.	Australasia	13,039,000
5.	Germany	9,236,000
6.	Spain	4,849,000
7.	Peru	2,514,000
8.	France	2,026,000
9.	Chili	1,910,000
10.	Austria	1,186,000
11.	Canada	1,158,000
12.	Japan	1,154,000
13.	Italy	1,154,000
14.	Colombia	1,123,000
15.	Central America	1,049,000
	All others	1,371,536
	Total	$118,748,546

LITERATURE.—Canadian Mining Manual, 1896. Quebec: Geol. Sur. IV. 1888 K. Ontario: Geol. Sur. III. 1887 H; Min. Res. Ont., 1890. British Columbia: Geol. Sur. III. 1887-88 R, IV. 1888-89 B: Rep. Min. of Mines, B.C., 1896, and Bull. No. 3.

LEAD.

By far the most important source of lead is the sulfid galenite (PbS), which frequently bears economic amounts of silver. It occurs either in granular or cubical crystals of a lead-grey color and brilliant metallic lustre. Cerussite, the carbonate ($PbCO_3$), containing 77 per cent. of lead, is white or grey in color, and of high specific gravity. Both minerals easily yield a malleable bead of lead before the blowpipe. The sulfate, anglesite ($PbSO_4$), also occurs, generally as a surface product of galena.

The ores of lead occur chiefly as mass deposits filling joints and irregular cavities in limestone. The lead has apparently been deposited with the sediments, and afterwards been brought in solution from the neighboring rocks into the cavities. Of this character are the deposits of Missouri, Iowa and Wisconsin. In Nevada the ore is frequently found at the contact of limestone with some dissimilar rock. In the gash veins of the limestone zinc is frequently found with the lead, sometimes one, sometimes the other, predominating, and silver being generally absent. A second occurrence of galena is in veins cutting ancient crystalline formations, as in British Columbia. These ores are more frequently silver-bearing, and that they are largely mined is shown by the fact that three-fourths of the lead produced in the United States is desilverized.

The great use of lead is in the manufacture of paint. Five-twelfths of the consumption of the United States in 1895 was used in the manufacture of white-lead. A considerable amount was also converted into litharge. Other uses are as leadpipe, shot, sheet lead, and in certain kinds of glass. Its alloys with tin, bismuth, antimony, are used as pewter, type and solder.

Spain, the United States, Germany and Mexico are the largest producers, and the United States, Great Britain and Germany the largest consumers

The total production of the world in 1894 was 617,000 metric tons, valued at three and a quarter cents a pound.

Canadian Localities and Production.—Galena occurs at a number of places in Nova Scotia in connection with Carboniferous limestone. At Smithfield, Colchester county, considerable development has been done on a large argentiferous deposit, and in Gloucester and Carleton counties, New Brunswick, some exploratory work was performed. In Quebec a very promising property of silver-lead has been developed on Lake Temiscamingue, and a few tons of argentiferous galena have been mined at the mouth of the Little Whale River, on the east coast of Hudson Bay. In Ontario silver-lead ores have been worked in Frontenac and neighboring counties, at Garden River, north of Lake Huron, and south of Lake Nipigon. These ores occur in veins, cutting Archæan schists or Cambrian argillites. In none of them were the silver contents high enough to make the properties successful at the present low value of lead. In British Columbia there are many deposits of galena rich in silver, and it is to that province that nearly the whole output of the Dominion is to be credited. The producing mines are in the Kootenay district, but many other promising localities are known.

THE MINERAL WEALTH OF CANADA.

PRODUCTION AND IMPORTS.

	1890.	1894.	1895.	1896.
Pounds of ore	113,000	5,703,000	23,076,000	24,200,000
Value	$5,800	$185,000	$750,000	$721,000
Imports unmanufactured	343,000	170,000	156,000	......
Imports manufactured	26,000	29,000	38,000	......

LITERATURE.—(See under Silver.) For British Columbia localities, see Geol. Sur. III. 1887-88, 155 R. Lake Temiscamingue: Geol. Sur. V. 1890-91, 90 S.

ZINC.

The most common zinc mineral is popularly known as blende or black jack, though mineralogists call it sphalerite. The first and last names refer to its blinding and deceiving or treacherous character, because, while at times resembling galena, it yields no lead, and because it occurs in all the colors of the rainbow. It has a peculiar resinous lustre, is scratched without difficulty with a knife, and is infusible before a blowpipe. In composition it is zinc sulfid, and when pure it contains 67 per cent. of zinc. The carbonate, smithsonite, results from the weathering of the sulfid, and is dirty white or brownish. Calamine, a silicate, is another zinc mineral often mined.

The ores of zinc closely resemble those of lead in their mode of occurrence and in their geological horizons, and often the two are intimately mixed. Blende, like galena, often carries silver, but it is more

THE MINERAL WEALTH OF CANADA. 91

difficult to part the silver and zinc than the silver and lead. Argentiferous blende occurs in the Thunder Bay district of Ontario and in the Kootenay district of British Columbia, but there is no production. A deposit of blende in Huronian diorite, north of Lake Superior, was exploited for a time, but operations have ceased. Kansas, Wisconsin, Missouri and New Jersey are the zinc-producing regions of this continent. Two-thirds of the ore of the world is mined in Germany; Italy is the second producer, followed by the United States and France. All of the Italian ore is exported, and Belgium, using imported ores, ranks second as a producer of metallic zinc, Germany having the first position. The total production of the world for 1894 was 383,225 metric tons, of which Canada took $130,000 worth, mostly manufactured.

CHAPTER VIII.

ARSENIC, ANTIMONY, TIN, ALUMINUM AND MERCURY.

Arsenic.—This element is little used in the metallic state, and then only as an alloy, the chief of which is with lead. Shot is hardened by the mixture of about forty pounds of arsenic with a ton of lead. Its most important use is in the manufacture of colors, particularly greens. Paris green is a commercial name for several chemical compounds used as colors, and also as insecticides. A small amount of the metal is used in making certain kinds of glass and in fireworks.

Arsenic is widely distributed in nature, occurring usually as a double sulfid and arsenid of iron, nickel or copper. Mispickel, or arsenopyrite (FeAsS) the chief mineral, is hard, brittle, silver-white, and gives a garlic odor when heated. Considerable deposits of it occur in Hastings county, Ontario, where it has been mined for the gold it contains. The output is, however, very irregular, in 1885 the product being valued at over $17,000, and in 1895 at nothing. Commercial arsenic has sold for some years at about four cents a pound, but in 1895 the price advanced to nine cents, and even at that figure it does not pay to produce the metal, except as a by-product. Cornwall and Devon, England, and Freiberg, Germany, supply the market with 7,000 to 9,000 tons a year. Canada imported in 1895 nearly 600 tons, valued at $32,000.

Antimony.—This metal frequently occurs as a mineralizing agent with ores of silver. The chief source is, however, the sulfid stibnite (Sb_2S_3), a soft lead-grey easily fusible mineral. It is recognized by the white fumes and odor of burning sulfur which it gives when heated with a blowpipe.

Stibnite has been mined at Rawdon, in Hants county, N.S., where in a gangue of quartz and calcite it occurs in a vein cutting Cambrian slates. The ore is of good quality, and in places is auriferous. At Prince William, York county, N.B., there are numerous large well-defined veins carrying quartz and stibnite in Cambro-Silurian slates. Several mining companies have operated there, reducing the ore in part and shipping the remainder to Massachusetts, where it was used in the manufacture of rubber. Ores of antimony have also been mined in South Ham, Wolfe county, Que. None of these properties are now in operation, litigation and the continually decreasing value of the product having forced them to close. Antimony, which was worth fifteen cents a pound in 1891, was quoted at seven cents in 1895. Antimony ores, probably in economic amounts, are reported from several localities in Ontario and British Columbia. In the latter province they are frequently argentiferous.

France is the largest producer of antimony, and Italy, Japan and New South Wales contend for second place. In 1893 the total production of ore was

15,000 tons, which would yield about 6,000 tons of antimony. In 1885 the Canadian product was 758 tons; in 1895 it was nothing. The imports in 1895 were forty tons, valued at $6,000. The great use of antimony is as an alloy with lead in making type metal.

Tin.—This is the only important metal of which no economic deposits occur in Canada, for, apart from a few mineralogical curiosities, it is unknown here. North America as a continent seems almost destitute of it, for in spite of very heavy protective duties the Americans have failed to develop any successful mines, though small amounts have been got in Dakota, California and Mexico.

The oxid of tin, cassiterite, is the only ore. The mineral is brown to black in color, of brilliant lustre when in crystals, hard and heavy. It is infusible before the blowpipe on charcoal, but with soda can be reduced to minute malleable beads of tin. Tin ore occurs in two ways: First and most important is the "stream tin," which is simply a placer deposit like that of gold, and due to the same cause, *i.e.*, to the weight of the mineral. These placer deposits are widely scattered over the world, but are comparatively rare. They are derived, of course, from veins which constitute the second class of deposits. Here the ore is disseminated in bunches and grains in the veins and in the ancient crystalline rocks, which they cut. Cornwall, England, is the most famous tin region of the world, though the original placers are exhausted and the veins themselves are not so productive as formerly. The ore is frequently found in a peculiar granite rock called greisen, which lacks felspar. Pyrite, chalcopyrite, blende, tourmaline, wolfram, topaz are often associated with the tin ore. Considering the immense granite areas in Canada, it would seem probable that tin will yet be discovered here.

The great use of tin is as a coating for iron in the manufacture of tin-plate. Small amounts are used in alloys, such as bronze, bell-metal and solder. In 1895 the production was about 80,000 long tons, of which 63,000 came from the Malay peninsula, and England, Tasmania and Bolivia produced nearly all the remainder. In 1895 the average price was fourteen cents a pound. Canadian imports average over one million dollars a year.

Aluminum is the most abundant metal in the earth's crust, and the third element in amount. It is found in hundreds of minerals, chiefly complex silicates like garnet, felspar and mica. Ordinary clay is a hydrous silicate of aluminum which, when pure, contains 21 per cent. of the metal. Notwithstanding the great number of minerals and their wide distribution, the ores of aluminum are very few. In most cases the chemical combination is too strong for profitable separation with our present methods. Corundum, the oxid, might be used, but it is too valuable as an abrasive to be employed as a source of the metal. Cryolite, a sodium aluminum fluorid, was until recently the chief source, the mineral being brought from Greenland. Bauxite, the mineral used at the present time, is a hydrated oxid of aluminum with iron replacing part of that metal. Silica, phosphoric acid, lime, and magnesia, are common impurities. In composition and mode of occurrence it resembles limonite. The mineral is white, yellow or red, soft and granular. It occurs in large amounts in France, Italy, Ireland, Georgia and Alabama, but is not yet known in Canada.

Bauxite is treated chemically and changed into the oxid of the metal (Al_2O_3), and this is reduced by a powerful electric current in a bath of molten cryolite. Only two companies are at present producers. One has works at Niagara Falls and Pittsburg, the other in Switzerland and France. The product in 1895 was nearly 1,300 tons, valued at 50 cents a pound. The demand for this metal will increase enormously once it can be marketed as cheaply as copper or zinc. In 1886 the price was $12.00; in 1892 it had fallen to 50 cents, and that seems to be the limit for the present.

Mercury.—The only ore of mercury is cinnabar, the sulfid (HgS), which contains when pure, about 87 per cent. of the metal. The mineral is bright red or brownish-red in color, is of high specific gravity, and is easily vaporized before the blowpipe. Often specks of the bright metal are scattered through the red mineral. It is found as an impregnation of various rocks which have been shattered and fissured by eruptive rocks, which are always found near at hand.

There are three important regions: Spain, where the cinnabar impregnates a sandstone of Silurian age; California, where the deposits are of Cretaceous and Tertiary age, and Austria, where the ore occurs in nearly vertical strata of Triassic age. The mineral seems to be the result of volcanic action, which has vaporized mercury, sulfur and steam at some distance below the surface. These vapors have then forced their way up through the shattered superincumbent rocks, and on cooling the mercury and sulfur have been united and deposited.

Around Kamloops Lake, British Columbia, a number of veins have been found in volcanic rocks of Tertiary age. Exploratory work has yielded good results, and a continuous output is promised.

The great use of mercury is in the recovery of gold and silver by the amalgamation process. As, however, the quicksilver can be used over and over, the market does not increase rapidly. Another important use is in the manufacture of vermilion paint. Small amounts are used in making mirrors, thermometers, barometers and medicinal compounds. The output in 1894 was 3,952 metric tons, of which two-fifths came from Spain and one-quarter from the United States, the remainder being furnished by Austria, Italy, Mexico and Russia.

LITERATURE.—Arsenic: Rothwell, "Mineral Industry," 1895; Min. Resources of Ontario, 1890; Bur. Mines, Ontario, 1893. Antimony: "Mineral Industry," 1895. Tin: "Min. Indus.," 1895; Louis and Phillips, "Ore Deposits." Aluminum: Richards, "Aluminum," 1890; "Min. Industry," 1892. Mercury: "Min. Industry," 1895; Rep. Min. Mines, B.C., 1896.

SECTION II.
MINERALS YIELDING NON-METALLIC PRODUCTS.

CHAPTER IX.

SALT, GYPSUM AND BARITE.

SALT.

Occurrence.—Common salt, so important to the welfare of the human race, is widely distributed, few countries being unable to supply themselves in case of need. Not only is the geographical distribution of large extent, but the geological horizons in which it is found are very numerous. Upper Silurian beds are found in Ontario and New York; Devonian ones in Manitoba and Athabasca; Lower Carboniferous salt springs are found in Cape Breton and New Brunswick, and beds of the same period in Michigan furnish much of the salt of the United States; Permian beds are found in Texas, and the famous deposit of Stassfurt, Germany, was laid down in the same period; in the Triassic beds are found the deposits of Kansas and Cheshire, England, and some salt springs on Vancouver Island come from the Cretaceous just above; in Tertiary times were deposited the great salt beds at Wieliczka, Austria, and some smaller ones in Louisiana. Even in historic times deposits have been formed in the arid regions of the west of North America.

Salt, known to mineralogists as halite, occurs in nature either in solid masses, known as rock salt, or in solution in water The solutions, or brines, are found (1) in oceans or salt lakes, (2) in salt springs, (3) in porous rocks, held in by impervious beds above and below. On drilling a hole through the upper retaining bed the third class may become the second.

Neither the rock salt nor the brines are pure as they occur in nature. The most common impurities are the sulfates of calcium, magnesium and sodium, the chlorids of calcium, magnesium and potassium, and the carbonates of calcium, magnesium and iron; clay, also, is found quite frequently in rock salt. The amount of the impurities is variable, but usually in salts of commercial value it is quite small. The following analyses show the composition of two standard natural salts:

	Goderich, Ont.	*Cheshire, Eng.*
Sodium chlorid, or salt	99.687	96.70
Calcium chlorid	.032	.68
Magnesium chlorid	.095	....
Calcium sulfate	.090	.25
Insoluble in water	.017	1.74
Moisture	.079	.63
	100.000	100.00
Total impurity	.234	2.67

Origin.—The sea has probably been salty since the time when the cooling earth first allowed the clouds of vapor to condense upon its surface. The hot, primeval ocean, under greater pressure than now, must have been a powerful solvent. No doubt its saltiness has been increased since then by the incessant and large contributions of every stream. Running water, as it percolates through our soils, dissolves out here and there grains of salt and gypsum and limestone, and hurries off with them to the ocean. The St. Lawrence, as it leaves Lake Ontario, carries one and a half tons of mineral matter every second to be deposited in the ocean and make it saltier. About 3.5 per cent. of ocean water consists of solids, of which common salt makes 2.7 per cent.; other constituents are magnesium chlorid, 0.4 per cent.; magnesium sulfate, 0.2 per cent., and twenty-three other elements.

Through changes of level and other causes, oceanic waters have been at times confined in lagoons, where, as evaporation went on, the calcium sulfate was first deposited as gypsum, and later, with greater concentration, the sodium chlorid was precipitated. Mixed with these were frequently marls and clays derived from erosion of the neighboring land. Last of all came the deposition of the potassium and magnesium salts as shown by the beds of Stassfurt, Germany. In many cases, however, the sea seems to have overleaped the boundary at intervals and furnished fresh solutions for second and third deposits. Only in a few cases have the more soluble salts of potassium and magnesium been deposited as at Stassfurt. The following section at Goderich, Ontario, shows six distinct beds of salt with intervening beds of marine-formed dolomites and marls:

Beginning at the Surface.	Feet.
Clay, gravel and boulders	79
Dolomite and limestone	797
Variegated marls with beds of dolomite	121
Rock salt, first bed	31
Dolomite with marls toward base	32
Rock salt, second bed	25
Dolomite	7
Rock salt, third bed	35
Marls with dolomite and anhydrite	81
Rock salt, fourth bed	15
Dolomite and anhydrite	7
Rock salt, fifth bed	14
Marls, soft, with anhydrite	135
Rock salt, sixth bed	6
Marls, dolomite, and anhydrite	132
	1,517

A total of 126 feet of rock salt.

In regions of great evaporation salt lakes are frequently found. Streams carry soluble salts from the land, and if the water is removed only by evaporation the closed basin becomes gradually saltier. The Great Salt Lake of Utah and the Dead Sea may thus ultimately become beds of rock salt. Salt springs are but mineral waters particularly rich in sodium chlorid, which derive their salts either from subterranean masses or from salts disseminated through clays and marls. These brines frequently collect in porous rocks and are often associated with petroleum and gas. In the opinion of Hunt the saline springs of the Palæozoic rocks of Ontario and Quebec derive their ingredients from the sea water held in the interstices of the marine sediments of the period.

Canadian Localities.—A number of salt springs arise from the Lower Carboniferous rocks of Nova Scotia and New Brunswick, but the proportion of salt is too small to be of economic value. About five hundred bushels are made annually at Sussex, N.B., which is used locally for table and dairy purposes.

In a belt of country ten to fifteen miles wide, and extending from the Niagara River to Southampton, Ont., rocks of the Onondaga period of the Upper Silurian form the outcrop, and these are overlaid to the south-west by Devonian strata. At numerous wells sunk through these overlying rocks for 1,000 to 2,000 feet, beds of salt have been found. The record of a boring for a Goderich well, given above, is typical. At first the salt was supposed to be confined to a limited area near Lake Huron, but it is now known to extend south through parts of Middlesex, Kent and Essex counties, as well as under South Bruce, Huron and Lambton. At Kincardine the salt bed is found 888 feet below the surface; to the south the depth increases, being 1,170 feet at Clinton and 1,620 at Courtright. Farther south, at Windsor, the upper salt bed rises to 1,272 feet. Salt from the same horizon is found across Lake Huron at St. Clair and Saginaw, but the brines which are evaporated at the latter place come from a higher horizon, that of the Lower Carboniferous.

The quantity of salt is inexhaustible. At Goderich the six beds aggregate 126 feet of solid salt, to say nothing of the quantity distributed through the marls. At Blyth a bed eighty feet thick is found; at Petrolia, one 105 feet thick; at Windsor the well is seventy-nine feet into the second bed without piercing it. All the beds are not of equal purity; the second and third at Goderich are among the purest known, yielding on analysis 99.7 per cent. of salt.

Numerous salt springs are found in the Devonian area to the west of Lake Winnipegosis, but no beds of rock salt have been discovered. These brines, though weak, have been used in the past as a source of salt. The process of manufacture as carried on by the Hudson's Bay Company was crude in the extreme. A hole five or six feet deep was made in the soil, and from this the water was ladled into kettles near at hand. From these the salt was scooped as it formed, and after draining for a short time was packed in birch bark for shipment. Farther to the north, along the Athabasca, similar springs are found, and have been used by the same company.

Manufacture.—Throughout the Goderich region the water that finds its way downward on the outside of the pipes which are sunk, forms an almost saturated solution, which is pumped to the surface and evaporated. A saturated brine contains 25.7 per cent. of salt; the brines of Ontario, twenty to twenty-four per cent., in which respect Canadian manufacturers

104   THE MINERAL WEALTH OF CANADA.

have a great advantage, those of Syracuse, N.Y., containing only eighteen to twenty per cent. In some cases water is forced down between an inner and an outer pipe and drawn up through the inner.

Evaporation of the brines is accomplished either by artificial heat, or by solar heat, or by congelation. Solar evaporation of ocean water is also practised in California, Scotland, etc. Congelation is practised in Norway. The ice which forms on a solution of salt consists of nearly pure water, and by repeated removal of the frozen surface a stronger brine is gradually obtained. In Ontario the brine is usually evaporated by artificial heat in iron pans one hundred to two hundred feet long and twenty-five wide.

Uses.—The chief use of salt is in seasoning and preserving foods, and as this depends on population there can be but a slow increase in production in Canada. Moreover, salt for use in the fisheries is imported free of duty, and as vesselmen carry it westward for almost nothing (it saves ballast), English salt can be sold in Montreal as cheaply as Canadian. Salt is, further, the basis of many important chemical industries, caustic soda, sodium carbonate, hydrochloric acid and bleaching powder being all derived from it. A small amount is used as a fertilizer and in the reduction of ores of silver.

SALT STATISTICS OF CANADA.

	1886.	1895.
Production, tons	62,359	52,000
Value	$227,000	$160,000
Exports	17,000	1,000
Imports paying duty—tons	6,133	4,200
" " " —value	$39,0 0	$30,000
Imports duty free—tons	90,103	101,000
" " " —value	$255,000	$333,000

LITERATURE.—Geological occurrence in Ont., Reports Geol. Sur., 1863-66, 1866, 1874-75, 1876-77; Occurrence, etc., in Man., Geol. Sur., V. 1890, pp. 219-224 E. Statistics, Geol. Sur. Rep. S; Min. Resources of Ont., 1890.

GYPSUM.

Gypsum ($CaSO_4 + 2aq$) is a soft mineral consisting of sulfate of calcium and water. It is usually white or grey in color, but may be red, brown, or black, if impure. It occurs at times in distinct plates, clear and transparent; again in fibres with a pearly lustre, giving rise to the name satin spar; more usually it is a massive, dull-colored rock, a fine-grained variety of which is known as alabaster.

Gypsum often forms extensive beds in stratified rocks, especially in limestones and calcareous shales, and occurs in all formations from the Silurian upwards. In Canada it is found in the Lower Silurian of Quebec, in the Onondaga division of the Upper Silurian in Ontario, and in the Lower Carboniferous of the Maritime Provinces. Large deposits were made in Triassic time in the western United States, and in Eocene time in Europe.

Canadian Localities.—Gypsum occurs in immense beds through the Lower Carboniferous strata of northern Nova Scotia. In Cumberland it outcrops along a line from Minudie to Wallace, particularly at Napan River and Pugwash. It is much more abundant in Hants and Colchester, particularly the former. Near Windsor there is found a "long range of cliffs of snowy whiteness," which, however, contain much anhydrite as well as gypsum. It is quarried for export at Windsor, Cheverie, Walton, Stewiacke and other places, with shipping facilities. The deposit is inexhaustible; the amount quarried is only limited by the demand. In Pictou a bed of economic value exists on the East River, but too far from navigation. Eastward the beds are found in Antigonish, where a cliff of gypsum, white and red, 200 feet in height, fronts the ocean. At Plaister Cove across the strait an enormous bed is found, two-thirds of which, however, is anhydrite. It is also found in Inverness, Victoria and Cape Breton counties. Nearly the whole product of Nova Scotia is shipped in the crude form to the eastern United States.

Gypsum, according to Dawson, "is a very abundant mineral in New Brunswick, the deposits being numerous, large, and in general of great purity. They occur in all parts of the Lower Carboniferous district, in Kings, Albert, Westmoreland and Victoria, especially in the vicinity of Sussex, in Upham, on the

North River in Westmoreland, at Martin Head on the Bay shore, on the Tobique River in cliffs over 100 feet high, and about the Albert Mines. At the last-named locality the mineral has been extensively quarried from beds about sixty feet in thickness, and calcined in large works at Hillsborough." At present the mineral is shipped from Albert and Victoria counties, most of it going in a crude condition to the United States and selling at about 90 cents a ton.

In the valley of the Grand River from near Cayuga to Paris, Ontario, for a distance of forty miles, gypsum frequently outcrops. The beds are lenticular in shape, the greatest diameter being about a quarter of a mile, and the thickness three to seven feet at the maximum, and nothing at the edges of the lenses. The beds are horizontal and are capped by thin bands of limestone and the drift, or by the latter alone, which gives the country a hummocky appearance. Some parts of the gypsum are grey, others white, the latter being purer and usually at the top. A large number of mines have been opened. Usually a level is run in from the valley of the river and the mineral brought out on a car. It is ground for land plaster and calcined to make plaster of Paris. The former finds a market in south-western Ontario; the latter, under the trade names of "Adamant Wall Plaster," "Alabastine," "Plastico," is sold throughout the Dominion. These deposits are found in the Onondaga formation of the Upper Silurian, which has been described earlier in the chapter as salt-bearing. It outcrops between Lakes Erie and Huron for a distance of 150 miles, and the gypsum-bearing area may yet be considerably extended.

Along the Moose River for a distance of seven miles banks of gypsum ten to twenty feet high have been found. Apparently these beds are Devonian. The deposit is, of course, too far away to be of any value. Gypsum is so widely distributed on this continent, and in such large amounts, that it cannot be shipped with profit to any long distance.

In northern Manitoba two beds, respectively twenty-two and ten feet in thickness, have been reported, and farther to the north-west along the Mackenzie River it has been found. On the Salmon River, British Columbia, it also occurs in economic amounts, but at none of these localities is it mined.

Origin.—A number of theories have been advanced to account for the great beds of gypsum. The one most commonly accepted is that given above in connection with the origin of the salt beds, viz., the evaporation in closed arms of the sea of salt water. Sediment would be deposited first, then gypsum; and as evaporation continued, salt would be precipitated. This is the normal order the world over, but every gypsum deposit has not of necessity an overlying salt bed, as evaporation frequently was not continued long enough; and in other cases water afterwards dissolved and carried off the salt which had been formed. Hunt has extended this theory somewhat. He holds that the sulfate of calcium in the sea water is due to a chemical reaction between bicarbonate of calcium and sulfate of magnesium, two soluble salts brought down from the land. Evaporation would cause the precipitation of gypsum followed by a hydrous carbonate of magnesium. If a calcium carbonate were also precipitated, it would mix with the magnesium salt, and on being slightly heated yield dolomite. Dana has supposed that the gypsum of Ontario and New York is due to the action of sulfuric acid springs on limestone, and that this might account for the mound-like appearance. Logan, however (Geol. Can., 1863, p. 352), thinks that the gypsum was formed at the same time as the shales that overlie it, and that the mounds are due to the removal of softer parts of the shales. Another theory which accounts for the mound-like deposits is that of hydration. Anhydrite ($CaSO_4$), which is gypsum without its water crystallization, is found in many sedimentary deposits, and as it is capable of taking up 25 per cent. of its weight of water, and of forming gypsum, but in doing so swells considerably, this would account for the dome-like masses. Dawson adopts the sulfuric acid theory to account for the immense deposits of Nova Scotia. He assumes that the acid given off by volcanoes found its way along the bed of the ocean, until it met with beds of calcareous matter which it changed into gypsum, and this agrees with the fact that gypsum is only found associated with marine limestones.

Uses.—Gypsum, ground to a fine powder, is used as a fertilizer. It is also ground and heated, when it loses its water of crystallization and becomes plaster of Paris. This substance has the valuable property of taking up the water again and hardening, so that it is used to form moulds, models and cornices. Tinted with proper materials it forms a beautiful decorative finish for walls, cheaper forms being even used as common wall plaster. The World's Fair buildings at Chicago owed their beauty to a white coating of stucco made from gypsum. Fine, granular, semi-transparent varieties known as alabaster are carved into ornaments.

STATISTICS, 1894.

	Tons.	Value.
PRODUCTION—		
Nova Scotia	168,000	$148,000
New Brunswick	53,000	48,000
Ontario	2,300	6,200
Total	223,300	$202,200
Exports	160,000	158,000
Imports, crude and manufactured	....	4,200

LITERATURE.—Localities: Nova Scotia and New Brunswick— Dawson, "Acad. Geol."; Ontario—Geol. Can., 1863; Min. Resources Ontario, 1890; Bur. Mines, 1891. Manitoba— Can. Rec. Sci., III. 353, 1889. North-West Territories—Geol. Sur., 1888, 30 D, 101 D. British Columbia—Geol. Sur., 1889, 42 S. Origin: Hunt, "Chem. and Geol. Essays," 1875, Chap. VIII.; Dawson, "Acad. Geol.," 1878, p. 262; Dana, "Geol.," 1895, p. 554. Production: Rep. S of Geol. Sur. Can.

BARITE.

Barite ($BaSO_4$) is connected chemically with gypsum and may be considered here. It is also known as barytes and as heavy spar. It is a common vein-stone especially with lead and zinc ores, and in Nova Scotia with iron ores. It also occurs as veins or pockets in limestone and sandstone, and these latter deposits are of greater commercial value since they are purer. It is widely distributed in Canada but only mined in a desultory way. At a number of points in Pictou and Colchester counties, N.S., as Hodson, Brookfield, Five Islands, it has been mined and exported, but the total production has been only a few thousand tons. A vein three feet wide at Hull, Que., is the source of a few tons of material used in Toronto. On McKellar's Island, Lake Superior there is a deposit of quartz, calcite and barite sixty feet in width. It is only mined intermittently, though one of the best deposits ever found.

The chief use of barite is as a pigment; for this purpose it is usually mixed with white lead, which it closely resembles in color and weight. By some it is considered an adulterant, though others claim that it gives greater body to the paint and that the mixture resists the action of the weather better than pure lead. Barite should be free from quartz grains and iron stains, though the latter may be removed by boiling with sulfuric acid. In 1894 the shipments were 1,080 tons, valued at $2,830.

CHAPTER X.

APATITE AND MICA.

APATITE (Gr. *απᾶτέ*, deception) occurs in green, red, blue, white, and even black crystals or crystalline masses, the former being hexagonal in outline and frequently of large size, one from Buckingham, Que., weighing 550 pounds and being seventy-two and a half inches in circumference. Apatite is mainly calcium phosphate, its composition being represented by the formula, $3Ca_3 2PO_4 + CaF_2$, though fluorin may be replaced by chlorin. An average of seven Canadian apatites, analyzed by Hoffman, shows calcium phosphate, 87.4 per cent.; calcium fluorid, 7.4 per cent.; calcium chlorid, 3.9 per cent.; calcium carbonate, 0.7 per cent.

Distribution.—Apatite is widely distributed, few igneous and metamorphic rocks being destitute of it, but the quantity is, in most cases, insignificant. The mineral in economic amounts has been found only in Canada, Norway and Spain, and there in the older rocks. In Canada it is found in two localities. One, in Ontario, stretches from a few miles north of Kingston one hundred miles in a northerly direction, and is fifty to seventy-five miles in width. The other, in Quebec, extends northward from Hull about sixty miles, and is fifteen or twenty miles in breadth. The latter, though smaller in area, has much richer deposits, and the chief mining operations centre there.

Occurrence.—In both districts the country rocks are gneisses and related rocks belonging to the upper part of the Lower Laurentian. For the most part they occur in belts with a north-east and south-west trend. Intrusive masses of pyroxenite occur in the country rock, the dikes sometimes running with the strike, at other times across it. As there are very seldom sharply defined walls, the pyroxene and gneiss shading into one another, some authors have held the pyroxenite to be a metamorphosed bed, but as the masses of pyroxene sometimes cut across the gneiss, this cannot be the case. The gneiss is frequently indistinctly stratified and often quite massive, and is usually more hornblendic in Ontario than in Quebec. The apatite deposits are usually found either in the pyroxenic or hornblendic rocks or quite near them. Sometimes the mineral is found in well-defined veins, but more usually it is in irregular masses throughout the pyroxenic rock, in some places apatite predominating, in others pyroxene, or mica, or felspar. The "pockets" vary from a fraction of an inch to many feet in diameter, and while there is a vast quantity of waste rock to be mined, it has been pretty well established that the deposits are continuous. Associated with the apatite are a large number of minerals, about thirty in all. Zircons, sphenes, scapolites and micas are found in almost unequalled size and perfection.

Origin.—Many diverse views are held concerning the origin of the Canadian apatites. Sir W. Dawson and others believe in an organic origin, and suppose that coprolites and phosphatic nodules of the original sediments have undergone metamorphism along with the muds and sands which held them, and so account for the bedded character of many of the deposits. Veins have probably been formed in some cases by subsequent segregation from these beds. Others hold that there is absolutely no evidence of organic origin. Selwyn, formerly director of the Geological Survey, asserts that "they are clearly connected for the most part with the basic eruptions of Archæan date." The same origin is held by the Norwegian geologists for the apatite deposits of their country, which are known to closely resemble those of Canada. The general view seems to be that the apatite and accompanying minerals have been segregated from the surrounding rocks into irregular masses without the existence of any true fissure.

Production.—Mining operations were begun in Ontario about 1850, but owing to the pockety character of the deposits were not vigorously prosecuted. Much of the ore was raised by the "contract system," farmers excavating pits a few feet deep, and, on exhausting a mass, opening another hole a little farther on. About 1871 extensive operations were undertaken in the Quebec district; drills were used for locating the deposits, and work prosecuted in a more systematic manner than had been the case in Ontario. Owing to the irregularity of the deposits

not more than 7 per cent. of the rock mined is apatite, but the mineral obtained is remarkably pure. The production is continually diminishing, as the following table shows:

1880	13,060 tons.
1885	28,969 "
1890	31,753 "
1891	23,588 "
1892	11,932 "
1893	8,198 "
1894	6,861 "
1895	1,822 "
1896	570 "

A few hundred tons are made into superphosphates at Smith's Falls, Ont., and Capleton, Que., and consumed locally, the rest being exported. About nine-tenths of the product comes from Quebec.

The principal market has been Great Britain, which, in 1891, imported 257,000 long tons of phosphates, of which Canada supplied about 8 per cent. The Canadian mineral is being driven out by the cheaper phosphates from the southern United States. Along the Atlantic coast from New Jersey to Texas are clays and marls carrying irregular nodules of phosphatic material varying from a grain to a ton in weight. In South Carolina and Florida rich deposits are found which are cheaply worked. Similar deposits are found in England, Belgium, France, Russia, and they are also reported as occurring in the Niobrara formation of Manitoba. Other competing products are guano, and basic slag derived from steel

116 THE MINERAL WEALTH OF CANADA.

works using an iron ore containing phosphorus. For the use of soluble phosphates as fertilizers, see the chapter on Soils.

LITERATURE.—Reports of the Geol. Sur., 1847-94, particularly those of 1873-74, 1876-77, 1877-78, 1888-89, pp. 89-111 K, 1890-91, pp. 153-161 S. For localities, see Min. Resources of Ontario, 1890, and pp. 108, 109 K Rep. Geol. Sur., 1888-89; "Bibliography," p. 110 K Rep. 1888-89, and Penrose, Bull. 46, U.S. Geol. Sur. There is a full survey of the phosphate of lime deposits of the world in Penrose's work. For economic details of working, etc., see Wyatt, "Phosphates of America."

MICA.

Occurrence.—On the failure of a profitable market for Canadian apatite, the producers of that mineral turned their attention to mica, which had until then been neglected. The old dumps of waste material were overhauled, the old workings re-examined, and new pits and trenches opened. Some phosphate is even mined now as a by-product of the mica industry.

The mica-producing territory embraces the two phosphate districts of Ontario and Quebec, and also some other localities. Loughboro' and North Burgess townships in Ontario, and Ottawa county in Quebec, are the chief seats of the industry. Commercial mica is further found in the Ottawa valley and Chicoutimi county, Quebec, and in Hastings county, Ontario.

Mica is found in very many kinds of rock, but usually in small flakes. Large plates are most commonly found in coarse granite, which occurs sometimes as dikes, sometimes without definite walls, as at the Smith and Lacey mine in Loughboro'. Here the most coarsely crystallized material has been excavated for a width of 15 feet to a depth of 130 feet. More frequently operations are confined to surface pits along the dike. The Villeneuve mine, Ottawa county, has a vein 140 feet wide, which is being worked to a width of 50 feet on the side of a hill. Here the felspar crystals are proving to be quite as valuable as the mica. Plates of the latter, measuring 30 x 22 inches, have been got from this property, and one crystal weighing 281 pounds yielded $500 of merchantable mica.

In mining care is taken to injure the mica crystals as little as possible by blasting. After being hoisted to the surface the mica is carried to the "stripping" room, where pieces of quartz, felspar, etc., are removed. Then in the "mica shop" it is split by knives to the required thickness, and afterward cut into standard shapes, which are put up in pound packages for shipment. There is great waste in cutting, one hundred pounds not yielding on the average more than ten of commercial mica.

The value of mica varies greatly, depending on the kind of mica and the size of sheet. For instance, the price list of the Villeneuve Mine, as cited by Obalski, quotes mica, 2 x 2 inches, at 50c. a pound, 4 x 4 at $9.10, 7 x 5 at $14.50. But this is for the white or muscovite mica. The amber-colored phlogopite and the dark biotite are not nearly so valuable. Rough, untrimmed mica, large enough to cut 1 x 3, is sold as low as 6c. a pound, and 4 x 6 at 60c. in ton lots. All three of these micas are silicates of aluminum with varying amounts of potassium, magnesium and iron.

Use and Production.—From the fact that mica is transparent to light and is not broken by heat or concussion it finds employment in stove panels, windows of men-of-war, eye-guards for foundry-men, etc. A recent use is as an insulator in electric machinery. For this it must be flexible, of uniform thickness and without small mineral crystals which conduct electricity, and for this purpose the dark varieties are as valuable as the light. Ground mica is used in making paint, as a boiler and pipe covering to prevent loss of heat, as a lubricant for heavy machinery, and for decorative effect in wall paper.

Canada, India, and the United States are the only producers. The amount mined in the last named country is decreasing, though the amount consumed is increasing. The production of Canada for 1895 was $65,000, that of the United States $38,000.

LITERATURE.—Min. Resources Ont., 1890; Rep. Geol. Sur., 1894, 73 S.

CHAPTER XI.

ASBESTOS, ACTINOLITE AND TALC.

"ONE of nature's most marvellous productions, asbestos is a physical paradox. It has been called a mineralogical vegetable; it is both fibrous and crystalline, elastic yet brittle; a floating stone, which can be as readily carded, spun, and woven into tissue as cotton or the finest silk." In Germany it is known as *steinflachs* (stone flax), and the miners of Quebec give it quite as expressive a name, *pierre à coton* (cotton stone).

The commercial substance includes a number of distinct minerals which are alike in being fibrous. The true asbestos of mineralogists embraces the fine fibrous forms of hornblende. The coarser fibres are known as tremolite or actinolite. All three consist of lime, magnesia and silica without water. The softer, silkier, and more flexible mineral which constitutes most of the commercial substance is chrysotile, a fibrous variety of serpentine, and chemically a hydrous magnesium silicate. Talc, steatite, or soapstone, also occurs in a fibrous, as well as in the usual massive form, and is very similar to chrysotile in composition and properties. The following table of approximate analyses will make these relations clear:

	Tremolite.	Actinolite.	Chrysotile.	Talc.
Lime	13	13	..	..
Magnesia	29	22	43	32
Iron oxid	..	7	..	..
Silica	58	58	44	63
Water	..	..	13	5

Quebec Asbestos Mines.—These mines, the most important source of asbestos known, yield 85 per cent. of the world's product, the only competing country being Italy, where the industry is declining. The asbestos is found in veins half an inch to six inches wide in masses of serpentine. The fibres are always at right angles to the sides of the veins, which are most irregularly distributed in the serpentine, cutting it in all directions and being badly faulted. The serpentine is associated with diorites which have been erupted through slates, or occasionally sandstones, of Lower Cambrian age. These serpentines extend from the Vermont boundary in a north-east direction almost to the extremity of Gaspé, and in three regions they have been found to contain asbestos. The first is near Mt. Albert in the Shickshock Mountains, where the mineral has not yet been found in economical amounts. The second is in Thetford and Coleraine, Megantic county; and the third district stretches from Danville through Orford and Bolton to the boundary.

Active mining is confined to the second district, and to Danville in the third. In the mines, which are in reality large open quarries, the serpentine is loosened

THE MINERAL WEALTH OF CANADA. 121

by blasting, hoisted to the surface, broken up, the refuse thrown on the dump, and the blocks bearing asbestos carried to the dressing or cobbing house. Here boys, with light hammers, separate the rock from the mineral and sort it into grades. At some mines elaborate machinery has been introduced for this purpose. The first grade contains the fibre over half an inch long well freed from rock. The "seconds" are poorer qualities of fibre, and the refuse makes "thirds." At the Thetford mines fifty to seventy per cent. of the output grades as "firsts," but at Black Lake the percentage is not so high. The intrusion of dikes of granite at the latter place seems to have caused sufficient heat to render parts of the asbestos harsher and less flexible. "Firsts" used to have a value of $125 to $150 a ton, and selected mineral even brought $250, but in 1895 $70 was an average price for "firsts."

The asbestos is derived directly from the serpentine in which it is found, and the latter is doubtless an alteration product of diorites rich in olivine. After the serpentines were fissured the veins were filled with material dissolved from the sides, and the crystals are accordingly always perpendicular to the walls.

In Ottawa county serpentine has been found in reticulated bands of varying widths in limestone of Laurentian age. In places it carries asbestos of good quality, from which a few tons have been brought as a test. Chrysotile has also been found in Hastings county, Ontario, and in the Fraser River valley, British Columbia.

Uses.—Chrysotile is flexible, non-combustible, and a non-conductor of heat and electricity, and on these properties its increasing use depends. It is spun into yarn, from which cloth is woven for drop-curtains in theatres, clothing for firemen, acid workers, etc. It is made into lamp-wicks, and gloves for stokers, and ropes for fire-escapes. It is felted into mill-board to be used as an insulator in dynamos, and as a fire-proof lining for floors. It is used to insulate electric wires, and as a covering to prevent loss of heat from steam pipes. It is a component of fire-proof paints and cements, and mixed with rubber it is used to pack steam joints. Indeed, one wonders how we ever did without it. Although Charlemagne is said to have had a table-cloth of asbestos which he was accustomed to cleanse by throwing in the fire, it was practically unknown until 1850. The Italian mineral was then experimented with, and some years later put on the market. In 1878 the first Canadian mine was opened, and the product steadily increased until 1890, when 9,860 tons, worth $1,260,000, were mined. There has since been a decline in value, the amount for 1896 being 12,200 tons, worth only $430,000. Little asbestos is manufactured in Canada, and consequently in 1894 we reimported goods to the value of $20,000.

LITERATURE.—Geological Occurrence, Localities, etc.: Reports Geol. Sur. I. 1885, 62 J; III. 1887, 106 K; IV. 1888, 139 K; V. 1890, 19 S; VII. 1894, 81 J. Methods of Mining, Cost, etc.: Report Geol. Sur. V. 1890, 12 SS. History and Uses: Jones' "Asbestos," 1888.

Actinolite.—This mineral occurs in several townships in Hastings and Addington counties, Ontario, in a band of serpentine, and is quarried in small quantities and ground in a mill at Bridgewater. The ground material retains its fibrous and flaky character, and mixed with pitch makes a strong and durable roofing material. Most of the product is shipped to Chicago.

Talc.—This is a soft mineral, white to green in color and with a greasy feel, which occurs in fibres, in foliated masses, and massive. The last variety is also known as steatite, soapstone and potstone. Some of the Indian's pipestone is likewise talc. The mineral is widely distributed in metamorphic rocks, especially the massive variety. It is found in the serpentine belt of Quebec described above; also in Hastings county, Ontario. Soapstone is unacted on by heat, and so is used to construct vessels exposed to high temperatures. Ground soapstone is used to fill paper, as paint, and as a lubricator. The compact mineral is used by tailors under the name of French chalk. Small quantities have been mined at Wolfestown, Quebec.

The fibrous form of talc is much rarer and also more valuable. Bands of it have been found in Addington county, which are said to compare favorably with the famous deposits at Gouverneur, N.Y., the production at which place in 1895 was worth $665,000. The talc is ground very fine, but still does not lose its fibrous character, and is then used in place of clay to give body and weight to paper, for which purpose it is better adapted than soapstone. The fibrous talc is also used as an adulterant in some asbestos manufactures. About 470 tons of soapstone, worth $2,138, were mined in 1895.

LITERATURE.—Quebec : Geol. Sur. IV. 1888, 151 K. Ontario: Rep. Bur. Mines, 1893.

CHAPTER XII.

PEAT, COAL, GRAPHITE.

PEAT.

IN all temperate and northern latitudes there are found areas of bog and swamp supporting a vigorous growth of moss. These mosses are mostly of the genus sphagnum, and characteristically grow upward as the lower parts die. Living in moist places as they do, these dead plants are immersed in water, and so preserved from rapid decomposition such as overtakes fallen forest trees. New vegetation springing up above gradually increases the pressure, and a slow carbonization results. In this way is produced a bed of vegetable matter slightly carbonized, retaining its fibrous structure and containing considerable water. The composition of this peat, after removal of the water, is about 60 per cent. carbon, 6 per cent. hydrogen and 34 per cent. oxygen. For comparison, the composition may be expressed in this way:

    Peat......Carbon, 100   Hydrogen, 10   Oxygen, 55
    Anthracite.   "   100   "   2.5   "   2

Often layers of marl are found at the bottom of the

peat, indicating that the deposit began in a fresh-water pond or lake, and that moss and rushes spreading out from the shores gradually filled up the basin. Successive layers are frequently found; beginning with the fresh-water shells, a layer of peat containing the remains of rushes and flags succeeds; then come layers containing mosses, and on top, after the bog is comparatively dry, heaths are associated with the sphagnums. Peat bogs grow upward at a rate varying from one foot in five years to one foot in twenty-five years or more.

Uses.—These peat bogs cover wide areas in the Old World, and are there used extensively for fuel. About one-tenth of Ireland is said to be covered with these deposits, and large areas exist in the continental countries. As the Irish bogs which are worked contain from eighty-eight to ninety-one per cent. of water, it is of course necessary to remove this injurious constituent. Three methods are available—exposure to air and sun, artificial heat or pressure. The peasants use the first, the others are used on a larger scale. Even then ten to thirty per cent. of water is present in the prepared turf used for domestic purposes in Ireland.

Peat is made into charcoal, of which it makes a useful variety. It is also distilled, yielding tar, oil, paraffin and illuminating gas. In New Brunswick and in Ontario companies are using peat to prepare "moss litter" as bedding for horses, etc.

Canadian Localities.—It would be useless to

attempt an enumeration of all the peat districts of the Dominion, so many are found. In general, it may be said that Anticosti Island, the east side of the St. Lawrence valley, the plain between the Ottawa and St. Lawrence, and the basin of the Moose contain extensive areas. Peat as fuel is only valuable where a cheap supply of coal is not available. For this reason the beds of Ontario and Quebec may become of economic importance. In 1874-75 33,000 tons of peat were made in Quebec and used on the Grand Trunk railway. Analysis showed approximately water 16 per cent., volatile matter 53, fixed carbon 24, and ash 7 for the manufactured article.

For details of Canadian beds, processes of manufacture, history of operations, consult Geol. of Can., 1863; Rep. Geol. Sur. IV., K 1888; Bureau of Mines, Ont., 1891, 1892.

COAL.

Coal is not a mineral in the strict sense of the word, for it is without definite composition. It consists mainly of oxygenated hydrocarbons with some simple hydrocarbons and free carbon. It may be defined as a "fossil fuel of a black color and strong consistency, which, when heated in closed vessels, is converted into coke with the escape of volatile liquids and gases." These oily substances are hydrocarbons mostly of the paraffin series. The varieties of coal depend on (1) the kind and the amount of the volatile ingredients, and (2) on physical characters, as structure, lustre, hardness.

Three chief varieties are usually distinguished and some hundred sub-varieties have been named : *Anthracite*, with a specific gravity of 1.35 to 1.8, bright lustre, and choncoidal fracture, has three to six per cent. of volatile matter, and burns with a feeble flame of pale color, does not smoke, and does not soften on being heated. It passes gradually through semi-anthracites into the second variety, *bituminous coal*. This includes a number of sub-varieties, all of which burn with a smoky flame, and give off oils or tar on distillation. In specific gravity they range from 1.14 to 1.40, and the volatile constituents may be as much as 66 per cent. Included here are (*a*) the caking coals, which soften on heating and are used to make coke; (*b*) the non-caking or free-burning coals, used for heating; (*c*) the cannel coals, particularly rich in hydrocarbons, and so of use in manufacturing coal gas. The third variety, called *Lignite*, has a specific gravity of 1.10 to 1.30, is usually dull brown in color, and frequently somewhat lamellar in structure. It is non-caking, rich in volatile matter, and usually has a large amount of water.

The following analyses compiled chiefly from the Geological Survey Reports will make the composition of the different varieties clearer. The results were obtained by fast coking.

128   THE MINERAL WEALTH OF CANADA.

Variety of Coal.	Locality.	Moisture.	Volatile matter.	Fixed carbon.	Ash.
Peat..........	Dismal Swamp, Va......	78.89	13.84	6.49	0.78
Peat, air dried.	St. Hubert, Que................	10.28	61.48	25.23	3.01
Cannel........	Crow Nest Pass, B.C...	2.10	57.71	30.33	9.86
Lignite	Souris River, Assa..............	13.94	41.92	38.35	5.79
Lignite	Swan River, Man................	9.66	43.16	43.61	3.57
Lignite	Moose River, Ont	11.74	41.39	44.03	2.84
Lignite	Edmonton, N. Saskatchewan	12.89	33.79	50.57	2.75
Bituminous....	Wellington Mine, Vancouver	2.75	38.03	52.64	6.58
Bituminous ...	Main seam, Sydney, C.B	.42	37.11	57.85	4.62
Bituminous ...	Main seam, Joggins, N.S	1.12	34.05	58.56	6.29
Bituminous ...	International Mine, Cape Breton.	0.87	35.41	58.56	5.16
Bituminous ...	Average Cumberland Co., N.S...	1.46	33.69	59.35	5.50
Bituminous ...	Average Vancouver Island	.....	30.33	60.23	9.44
Bituminous ...	Main seam, Pictou, N.S....... ...	1.55	27.99	60.84	9.62
Bituminous ...	Crow Nest Pass, B.C............	1.79	33.04	61.55	3.62
Bituminous ...	Comox Union Mine, Vancouver	1.34	30.01	65.82	2.83
Semi-anthracite	Bow River Pass, Ala............	0.71	10.79	80.93	7.57
Anthracite	Graham Island, B.C..	1.99	7.65	80.62	9.74
Anthracite	Mammoth vein, Pennsylvania....	3.42	4.38	83.27	8.20
Anthracite ...	Graham Island, B.C	1.89	4.77	85.76	6.69
Semi-anthracite	Bow River Pass, Ala.............	1.04	9.15	87.18	2.63
Anthracite	Pennsylvania	2.93	4.29	88.18	4.04

The following table of ultimate analyses shows the amount of each element present:

Variety of Coal.	Locality.	Carbon.	Hydrogen.	Oxygen and Nitrogen.	Sulfur.	Ash.	Hygroscopic water.
Lignite......	Medicine Hat, Assa..	54.35	3.34	17.52	0.17	7.30	16.82
Lignitic coal .	Belly River, Ala	62.39	3.99	16.82	0.77	6.85	9.18
Bituminous ..	Old Man River, Ala .	71.11	5.04	11.63	0.66	9.20	2.36
Bituminous ..	WellingtonMine,B.C.	72.65	4.89	12.77	0.36	6.58	2.75
Bituminous ..	Crow Nest Pass, B.C.	80.51	5.20	8.37	0.51	3.62	1.79
Anthracite ..	South Wales........	92.56	3.33	2.53	..	1.58	..

Impurities in Coal.—Carbon and hydrogen are the valuable constituents of coal. Nitrogen, oxygen and the mineral ingredients known as ash, are not deleterious except so far as they replace more valuable elements. Hygroscopic water which, on burning the coal, must be converted into steam, lessens the heating value of the fuel. Sulfur and phosphorus burn to offensive gases and act injuriously on iron, so that coals containing them are not suitable for domestic or smelting purposes.

The amount of ash in good coals varies from two to ten per cent. From the method of formation it is naturally somewhat larger in anthracite than in bituminous coal. In the best coal it does not seem to be greater than the amount of ash in the plants from which it is derived; but fragments of shale are usually present and increase the amount. Silica, alumina, lime, iron, potash and soda are the chief constituents of the ash.

Geological Occurrence. — Coal occurs in beds interstratified with shales, sandstones, fire-clays and limestones, the seams varying from a fraction of an inch to many feet in thickness. The "Mammoth" vein of Pennsylvania reaches a maximum of 50 feet, and the chief seam at Pictou, N.S., is 38 feet in thickness. These thick seams are not, however, all coal, for there are frequent partings of bituminous shale. The following section slightly condensed from Dawson's "Acadian Geology," shows the structure of the main seam at Pictou :

		Feet.	Inches.
1.	Roof shale	0	3
2.	Coal with shaly bands	0	6½
3.	Coal, laminated; layers of mineral charcoal and bright coal; band of ironstone balls in bottom.	2	0
4.	Coal, fine, cubical and laminated; much mineral charcoal	3	2
5.	Carbonaceous shale and ironstone, with layer of coarse coal	0	4½
6.	Coal, laminated and cubical	9	3
7.	Ironstone and carbonaceous shale	0	8
8.	Coal, with ironstone balls in bottom	1	2
9.	Coal	6	7
10.	Ironstone and pyrites	0	3
11.	Coal	10	3
12.	Coal coarse, layers of bituminous shale and pyrites	1	0
13.	Coal, laminated	2	1
14.	Coal with shale	2	3
15.	Underclay	0	10
	Thickness perpendicular to horizon	40	8
	Actual thickness	38	6

The beds occur for the most part in trough-shaped basins, and the different strata and coal seams are fairly persistent in arrangement and thickness over considerable areas. The Pittsburg seam of the Appalachian coal field underlies an area of 22,500 square miles. Compared with this the Canadian coal fields are of small extent, but the beds are frequently found throughout the whole field.

Below the coal seam there is nearly always a bed of clay, supposed to be the soil on which grew the vegetation that was subsequently transformed into the coal. Fossil roots, known as stigmariæ, are frequently found in these strata. The clays are often of great purity, and frequently are very refractory. Of course such clays, at the time they supported plant life, must have been horizontal; though now they, and the coal seams above, are frequently found highly inclined, as in the Pictou field. In the foldings to which the coal has been subjected it has in many cases suffered change. In the Bow River region of Alberta the coals of the plains are lignite; but as the mountains are approached the lignites are replaced by bituminous coals, and these in the Cascade basin in the mountains are replaced by semi-anthracites and anthracites.

Thin seams of coal have been found in the Silurian and Devonian systems, but none are of economic importance. The Carboniferous, especially the upper portions, is, in the extent and quality of its coal beds, by far the most important coal-bearing system. The Permian, Triassic, Jurassic, Cretaceous, Eocene, Miocene and Pliocene systems all contain coal, usually in small amounts and of poor quality. The Cretaceous and Tertiary coal-beds are often, however, of enormous extent, and some of the beds are of excellent quality.

Origin of Coal.—That coal is of vegetable origin is attested by the fact that the woody structure is still to be seen in some cases, and the microscope shows the cells of the original plant in many more. Spores of lycopods are recognized in some coals, and tree-trunks standing at right angles to the coal seam, are frequently found with their roots penetrating the clays below. The Nova Scotia beds have furnished many fine examples of these erect trunks.

These vegetable remains slowly lost their excess of hydrogen and oxygen, probably much as charcoal is at the present time made from wood, *i.e.*, by heating where no air is present. In this way the oxygen unites with a small part of the carbon and passes off as carbon dioxid, and a part of the hydrogen disappears as water. The following table, compiled from Thorpe, shows the gradual passage from wood to anthracite coal :

	Carbon.	Hydrogen.	Oxygen and Nitrogen.	Ash.
Mean composition of wood	49.6	6.1	43.1	1.2
Club-moss without ash	49.3	6.5	44.2	..
Humus, mean composition.....	54.8	4.8	40.4	..
Peat, Devon..................	59.7	5.9	34.4	..
Lignite, Cologne..............	67.0	5.3	27.7	..
Brown coal, Tasmania	71.9	5.6	22.5	..
Bituminous coal, Dudley......	79.7	5.4	14.9	..
" " Newcastle ..	87.9	5.3	6.8	..
Anthracite, Wales	93.5	3.4	3.1	..
" Peru...............	97.3	1.7	1.0	..

The club-mosses are the nearest living representatives of the coal vegetation, and the first two analyses show the great similarity in composition of very different plants. The oxygen and nitrogen are gradually eliminated, leaving a product each time richer in carbon. Apparently the hydrogen is not affected, but if a constant quantity of carbon is taken it, too, is shown to be given off.

	Weight of a cubic foot in lb.	Carbon.	Hydrogen.	Oxygen and Nitrogen.
Wood, average	30	100	12.3	86.8
Peat, "	50	100	9.7	54.7
Lignite, "	70	100	8.3	40.0
Brown coal, average	75	100	7.4	29.7
Bituminous coal, "	80	100	6.4	13.4
Anthracite coal, "	90	100	2.6	2.3

Wood exposed to the air quickly rots, and all the carbon is consumed, but below water the action goes on much slower, since little oxygen is present. In this way plant remains might be preserved for years, new accumulations but serving the better to prevent the oxidation of the carbon of the old. Noticing the gradual passage in composition and physical characters from peat to coal, it is but natural to suppose a peat bog to be the origin of all coal beds. Doubtless this peat bog theory is true for some of the lignite formations, but in the main it is incorrect. As the shales and limestones above and below the coal seams contain marine or brackish-water fossils, the beds must have been made in or near salt water. Nor have they arisen through the drifting of timber to the mouth of a stream and the silting over of the vegetable matter. This estuary theory does not account for fragile fern impressions and erect tree trunks and the stigmariæ in the under clay. Probably the vegetation flourished in swamps of brackish water along the coast and barely above sea level. After years of growth and decay a bed of vegetable matter was formed, and by a change of level the sea flowed over it, muds or sands were deposited and a slight elevation taking place a new growth of plants began. This in its turn was covered by the sea and a marine sediment deposited. And so by alternate risings and fallings of the land, by alternate marsh and sea, vegetable and mineral beds were deposited. The organic material under pressure slowly lost its gases and became coal, the variety depending on the age of the beds and on the amount of pressure and heat. Graphite, almost pure carbon, has originated, in some cases at least, through excessive heat and pressure applied to anthracite, and it seems to be the last stage in the progressive change from wood to carbon.

The Coal Fields of Canada.—*The Maritime Provinces.*—Throughout Nova Scotia and New Brunswick coal is found in rocks of the Carboniferous era, which are widely distributed and in places are of great thickness. Sir William Logan's section at the Joggins has a measured thickness of 14,570 feet, and the lowest part of the system is absent. Sir W. Dawson assigns a thickness of 16,000 feet to the Carboniferous of Pictou. The New Brunswick beds are very much thinner, 600 feet being about the average. Carboniferous rocks are exposed over about two-thirds of New Brunswick and one-third of Nova Scotia.

They border the Gulf of St. Lawrence from Gaspé through New Brunswick, northern Nova Scotia, including Cape Breton Island and western Newfoundland. Although of large extent, but a small portion of this area is coal-producing. Three fields are of economic importance, viz., Cumberland, Pictou and Cape Breton counties in Nova Scotia. Coal is found in other districts, but in too narrow seams to be of much value. A small amount is mined yearly in the vicinity of Grand Lake, N.B., but operations are of a desultory character.

The Sydney or Cape Breton field, which has been worked for almost two hundred years, extends from Miré Bay to Cape Dauphin, thirty-two miles along the north-east coast of the island. The land area of the coal measures proper embraces sixty square miles, and it has been estimated that within three miles of the shore two billion tons of submarine coal are available. If the millstone grit, which carries workable seams in places, is included, the land area of workable coal becomes 200 square miles. The field is divided into four basins by anticlinals, but the beds and coal seams are remarkably uniform for the whole district. Conglomerate followed by limestone constitutes the lowest 4,600 feet of the Carboniferous rocks. Next above is 4,000 feet of millstone grit. Succeeding this are the productive coal measures which include argillaceous and arenaceous shales, marls, underclays, limestones, black shales and coal. The measures are 1,850 feet thick, and of these forty to fifty feet are coal. The average number of seams is said to be twenty-four, of which six are three feet and over. Underclays are always present, and sandstone frequently covers the coal seam. The coal, which is all bituminous, is said to be more combustible than that of Pictou, and contains less ash and more sulfur. About a dozen collieries are being worked in this field.

The Cumberland county area has in general a trough-like structure, the rocks outcropping on the north dipping to the south, and those occurring on the north flank of the Cobequid Mountains dipping to the north. Cliffs in this county fronting Chignecto Bay furnish one of the finest sections of carboniferous rocks in the world. The famous South Joggins section exhibits almost a continuous series of beds 14,500 feet in thickness. The beds dip S. 25° W. at an angle of 19° and are exposed for about ten miles. They are made up of sandstones, conglomerates, shales, limestones and underclays filled with stigmariæ, the series containing no less than seventy-six coal seams each indicating a period of quiescence and a luxurious marsh. The thickest seam is, however, only five feet, and this has from one to twelve inches of clay along the middle. A number of collieries are operating here and on the continuation of these seams to the east. At Springhill the most productive colliery in the province is working in a distinct basin where there are five seams ranging from four to thirteen feet in thickness.

The Pictou field is a continuation to the east of the Cumberland carboniferous deposits. The thickness and number of the coal seams in parts of the dis-

THE MINERAL WEALTH OF CANADA. 137

trict are very remarkable. A part of the section at the Albion mines is given by Dawson as follows:

	Feet.	Inches.
Main coal seam (greatest thickness)	39	11
Sandstone, shale and ironstone	157	7
Deep coal seam	24	9
Shales, sandstone and ironstone, with several thin coals, viz., the Third seam, "Purvis" seam and "Fleming" seam, in all about twelve feet thick.	280	0
"McGregor" coal seam	11	0
Total	513	3

Here we have five seams aggregating nearly eighty-eight feet of coal in a distance of 513 feet. It is to be noted, however, that the measurements were made perpendicular to the surface, and that the beds are inclined at an angle of 20°. The main seam has an average thickness of thirty-eight feet, and at least twenty-four feet of this is marketable coal. Dawson calculates that this seam should yield 23,000,000 tons to the square mile, and other seams in the district half as much. Nothing like this amount is, however, attained in practice, the two main seams yielding about 10,000 tons to the acre, or 6,400,000 tons to the square mile. There are several reasons for this shrinkage: the district is badly faulted; the beds are steeply inclined, and so, besides being hard to work, soon reach unworkable depths; there are very sudden changes in the character of the coal, often making it worthless.

The coals of this field are non-caking and good steam producers, and some make good coke for iron furnaces. Their worst defect is the large amount of ashes which they contain.

For details of the geology of these fields and of the mines, consult the following:

Cumberland Co.: Dawson, "Acadian Geology," "Rep. Geol. Sur.," 1873-74, 1884-85, S 1886 to S 1892. Pictou Co.: "Rep. Geol. Sur.," 1866-69, 1890-91, S 1886 to S 1894. Poole, "Trans. N.S. Inst. Sci.," II. 1, 1892-93. Dawson, "Acad. Geol." Cape Breton Co.: "Rep. Geol. Sur.," 1872-73, 1874-75, 1882-84, S 1886 to S 1892. Fletcher, "Trans. Min. Soc., N.S.," III. 1894-95. Dawson, "Acad. Geol." Routledge, "Trans. Am. Inst. Min. Eng.," XIV. 542. Much information concerning all of them will be found in the annual reports of the Department of Mines of Nova Scotia.

Manitoba and the North-West Territories.— Throughout the plain region of Canada there is an immense tract of territory bearing coal. While the Carboniferous was the coal-forming era of the east, rocks of this age are destitute of coal in the west, their place being taken by the Cretaceous and early Tertiary formations. The coals vary from poor lignites to good anthracites, the quality improving as the mountains are approached. The most easterly beds occur in the Laramie formation in the Turtle Mountain district, Manitoba, where a bed is found some four feet thick and fairly persistent throughout the district. Throughout southern Assiniboia there is an immense area of Laramie rocks, carrying lignite in many places. In the Souris River valley Selwyn estimates that there is an area of 120 square miles, carrying 7,137,000 tons to the mile. These coals contain a large amount of water, and easily disintegrate on exposure, so that they are unsuited for transportation, but can be used locally.

THE MINERAL WEALTH OF CANADA. 139

Natural sections occurring on the river banks show seams of coal at scores of places throughout the Cretaceous and Laramie of south-western Assiniboia, and the whole of Alberta. At Medicine Hat, on the South Saskatchewan, in a bank 260 feet high, there are nine beds, aggregating sixteen feet of coal, two of these beds being each about five feet thick. At Coal Banks on the Belly River there are five seams in 42 feet. No data are available for estimating the exact extent and value of these enormous beds. Dawson has shown that in several districts in the Bow and Belly River valleys there are 5,000,000 tons to the square mile. The great seam on the North Saskatchewan maintains a thickness of 25 feet for three miles, and has been traced for 180 miles.

In the region of the plains the coal is lignitic, but superior to that widely used in Germany and Austria. In the foot-hills and in the isolated Laramie and Cretaceous basins of the mountains, the coal is bituminous. In one basin—the Cascade—pressure has been greater and an anthracite has been produced. This basin is 65 miles long and about 2 wide. The rocks, which are 5,000 feet in thickness, are shales and sandstones of the Kootanie division of the Cretaceous. Two seams of workable coal are here yielding the only anthracite produced in Canada. Outcrops of lignite are found in the river valleys far to the north. Coal beds on the Mackenzie River in latitude 67° N., and on the Lewes, a tributary of the Yukon, and elsewhere, may yet prove of great value.

For details see the following Geol. Sur. Reports: Souris River, etc , 1879-80 A ; Bow and Belly Rivers,

1880-82 B, 1882-84 C; Cascade basin, 1885 B; Analyses, physical characters, fuel value, 1882-84 M, 1885 M, 1887 T, 1888 R; Localities, Catalogue Section I. of the Museum.

British Columbia.—Coal was discovered in British Columbia in 1835, and a few tons mined each year until 1852, when operations were begun on a larger scale. Up to the end of 1896 over 11,000,000 tons have been mined, and the industry is growing continuously. Coal is found in two geological formations, the Mesozoic and Tertiary. Carboniferous rocks, though found in British Columbia, and often of great thickness, are never coal-bearing. The coal found varies all the way from a poor lignite, though first-class bituminous coal to a good anthracite.

The Cretaceous was the coal-bearing era in the province, and two periods of growth are recognized. The older is represented by the coal measures of Queen Charlotte Islands, Quatsino Sound, Vancouver Island, and Crow's Nest Pass in the Rocky Mountains. The upper coal measures of the Cretaceous are found at Nanaimo, Comox and Suquash on Vancouver Island. On the Queen Charlotte Islands both anthracite and bituminous coal are found. The beds in which the former is found are almost vertical, a fact connected with the metamorphism which the coal has undergone. Mining operations have been attempted on a bed six feet thick, but the difficulty of following the seams, the coal often being in a crushed and pulverulent state, has been a barrier, so far, to success. Valuable seams of bituminous coal, eighteen feet thick, are found in these islands. In the same horizon in the Crow's Nest Pass twenty seams of bituminous coal are reported, three of them being respectively 15, 20 and 30 feet thick, in all 132 feet. The area of this field is at least 144 square miles, and it promises to be one of the most productive fields in the Dominion. Selwyn calculates that there are 50,000,000 tons to the square mile.

The chief productive measures at present are in the upper portion of the Cretaceous system. This formation extends as a synclinal trough for 130 miles, the western side of the trough forming the eastern slope of Vancouver Island, and the remainder being under water. It is divided into two districts, the northern one, the Comox field, having an area of three hundred square miles, and the Nanaimo one to the south an area of two hundred square miles. At Comox the coal measures are 740 feet in thickness, and contain nine seams aggregating 16 feet of coal. The lowest and thickest averages 7 feet. At the Union mine in 122 feet, only a small part of the productive measures, there are ten seams with an aggregate of 30 feet of coal, the thickest bed being 10 feet. Richardson has calculated that in this field there are 16,000,000 tons of coal to the square mile. In the Nanaimo field there are two seams of workable coal, six to ten feet in thickness. The coal from both these fields is of excellent quality and much superior to the lignites found in Washington and Oregon States to the south.

The fuels of the Tertiary in British Columbia are usually lignites, though occasionally a bituminous coal is found. Most of them are found in rocks of the Miocene era, though at the mouth of the Fraser an area of eighteen thousand square miles is underlaid by the Laramie formation, which is a continuation of the lignite-bearing formation of Washington State. About twelve thousand square miles of igneous Tertiary rocks in the interior plateau are underlaid by sedimentary rocks of the same era, and these probably contain deposits of lignite in many places. Such beds have indeed been found and worked in the valleys of the Nicola and Thompson rivers. Many other localities are reported, a complete list of which is given by Dawson, Rep. R Geol. Sur. Can., 1887-88, p. 145. For details of the coal districts see the Geol. Sur. Reports, 1871-72, 1872-73, 1873-74, 1876-77, 1878-79, B 1885, B 1886, R 1887, A 1891, and the annual reports of the Minister of Mines of British Columbia.

Foreign Coal Fields.—In the United States there are several areas of Carboniferous coal, the most important one being that of Pennsylvania-Arkansas. The productive measures of this area are divided into three parts, viz., Appalachian, Illinois and Mississippi. Throughout the Western, Rocky Mountain and Pacific States there are immense areas of Cretaceous and Tertiary coals. Most of these are lignites, but some are good bituminous coals. The Atlantic and Pacific coasts of the United States are without good coal; the interior is well supplied. There are about 300,000 square miles of coal-bearing strata, but not more than 50,000 square miles are of economic importance.

THE MINERAL WEALTH OF CANADA. 143

In Great Britain the area of the coal measures is 12,000 square miles, the thickness being greater than in any other part of Europe. In France there is an area of 2,000 square miles; in Spain, 4,000; in Belgium, 518; in Austria, 1,800; in Germany, 1,700. In Russia there is an area of 30,000 square miles, but in not more than 11,000 are the beds of economic value. In China, India and Australia there are large areas of Permian age, and in Austria and Germany there are large areas of lignite in beds of Miocene age.

Production. — The following tables are self-explanatory :

ANNUAL PRODUCTION OF CANADA.

Year.	N. B.	N. S.	N. W. T.	B. C.	Total Tons.	Total Value.
1885	7,000(?)	1,514,000	40,000	366,000	1,880,000	$3,817,000
1894	6,000	2,527,000	200,000	1,135,000	3,868,000	8,499,000
1895	9,000	2,266,000	186,000	1,052,000	3,513,000	7,727,000
1896	....	......	......	......	3,743,000	8,006,000

IMPORTS AND EXPORTS OF CANADA.

		Imports.		Exports.	
		Tons.	Value.	Tons.	Value.
1894	Bituminous coal..	1,360,000	$3,315,000	1,104,000	$3,542,000
	Anthracite	1,531,000	6,354,000	......	......
	Coal dust........	118,000	50,000	......	......
1895	Bituminous coal ..	1,445,000	3,321,000	1,011,000	3,318,000
	Anthracite	1,404,000	5,351,000	......	......
	Coal dust	181,000	52,000	......	......

PRODUCTION OF COAL IN THE WORLD.

(From "Rothwell's Mineral Industry.")

(Metric tons, 2,204 lbs.)

Country.	1895.
Great Britain	194,351,000
United States	177,596,000
Germany	103,877,000
France	28,236,000
Austria	27,250,000
Belgium	20,415,000
Russia	7,551,000
Australia	3,975,000
Japan	3,650,000
Canada	3,187,000
India	2,650,000
All other countries	5,267,000
Total	578,209,000

LITERATURE.—"Reports Pennsylvania Geological Survey;" Dawson, "Acadian Geology;" Green, etc., "Coal, Its History and Uses;" Dana's "Geology;" Geikie's "Geology;" Details of equipment and production of Canadian mines in statistical reports (S) of the Geol. Sur., and in the annual reports of the Departments of Mines for N.S. and B.C.; also in Can. Mining Manual, 1896.

GRAPHITE.

Graphite is a soft greyish-black mineral, with a greasy feeling, consisting entirely of carbon. It is known also as plumbago and as black-lead, but both are misnomers since it does not contain that metal. Sometimes it occurs in hexagonal crystals, more usually in a massive state, either foliated, columnar, or scaly. It is found in beds or disseminated masses in metamorphic rocks as gneiss and crystalline limestone. In some cases it has certainly resulted from the alteration of coal by heat, occasioned by mountain folding, as in Rhode Island, or by the heat of errupted dikes, as in Texas. Some have held that all the graphite of the older rocks is of this origin, and that the immense deposits in the Laurentian gneisses are but the metamorphosed vegetable remains of that distant time. Of this we have no direct proof, and the absence of all fossil remains rather speaks against the theory.

Occurrence—Graphite is distributed through the older rocks in all parts of the world. It occurs in immense quantities of exceptional purity in the island of Ceylon, and it is from there that most of the commercial supply is now brought. Austria, Germany and the United States contain large deposits, and a considerable amount is mined yearly in these countries.

In Canada graphite is found in economic deposits in three localities. In the neighborhood of St. John, N.B., beds of argillites and limestones contain large quantities of disseminated graphite. Argenteuil and Ottawa counties, Que., and the line of the Kingston and Pembroke Railway, Ont., are the two other localities which are, however, geologically one. The Quebec region is the more important, and from it nearly all the mineral produced in Canada has come. According to Vennor the graphite is here found "in three distinct forms: 1, as disseminated scales, or plates in the limestones, gneisses, pyroxenites and quartzites, and even in some of the iron ores, as at Hull; 2, as lenticular or disseminated masses, embedded in the limestone, or at the junction of these and the adjoining gneiss and pyroxenite; and 3, in the form of true fissure veins, cutting the enclosed strata." The first method of occurrence is of the most importance economically, twenty to thirty per cent. of the rock frequently being graphite. The veins vary from an inch to two feet in width and contain the purest mineral. The rock is crushed and washed and the lighter graphite separated, the dressed graphite resulting containing three to ten per cent of ash, which by treatment with hydrochloric acid is easily removed. Hoffman has shown that so treated Canadian graphite is quite as pure and quite as incombustible as the Ceylon product. Vein graphite from Ceylon and Canada are almost identical, as the following analyses show:

Canada: carbon, 99.81; ash, 0.08; volatile matter, 0.11
Ceylon: "   99.79; " 0.05;   "   "   0.16

Notwithstanding these large and pure deposits the production of Canadian graphite is decreasing, the reason assigned being the lack of uniformity in the article put on the market.

Uses.—The uses of graphite depend on its infusibility, softness, and ability to conduct heat and electricity. One-third of the product is employed in refractory articles, as crucibles, furnaces, etc. It is a striking fact, illustrating the influence of the arrangement of the molecules of a substance on its properties, that we use pure carbon as charcoal or coke to heat our furnaces, and pure carbon mixed with fire-clay to make crucibles to resist the heat. Other uses of graphite are for stove polish, foundry facings, glazing powder, lubricating heavy machinery, electrotyping and pencil leads.

The production in 1895 was 220 tons, valued at $6,100, and of this 54 tons valued at $4,800 were exported. There were imported the same year plumbago manufactures to the value of $38,000.

LITERATURE.—Rep. Geol. Sur., 1873-74, 1876-77, 1888 K, 1890-91 S and S S.

CHAPTER XIII.

THE HYDROCARBONS.

PETROLEUM.

PETROLEUM is an oily liquid of disagreeable odor, usually greenish-brown in color but varying widely. In specific gravity it ranges from 0.6 to 0.9, some kinds being thin and flowing whilst others are thick and viscous. On the one hand, it graduates through maltha into asphalt or solid bitumen; on the other into natural gas. None of these substances are properly minerals. They are indefinite mixtures of a number of hydrocarbon compounds, chiefly of the paraffin series (C_nH_{2n+2}). The olefins (C_nH_{2n}) and benzenes (C_nH_{2n-6}) are present in small amount. The higher the value of n the higher the melting and boiling points, so that certain mixtures are gases, others liquid oils, and a third division are solids. The solid paraffins are soluble in the liquid ones, so that crude petroleum often yields large amounts of paraffin wax. This is especially true of the Ontario oil. The different liquid compounds are separated by distillation, and the crude oil is made to yield gasoline, benzine, naphtha, kerosene, lubricating oil, etc.

Occurrence.—Petroleum occurs in all the sedimentary formations from the Cambrian period to the present. Its geographical distribution is world-wide, but it is in comparatively few localities that it exists in economic quantities. It is associated usually with argillaceous shales and sandstones, and not infrequently is found impregnating limestones. Where these oleiferous rocks outcrop, the water of the wells and rivers frequently has a scum of oil. More often, and especially with the richer deposits, the oil beds are at some distance below the surface and covered with an impervious layer of rock. The source of the oil is undoubtedly the animals and plants which were entombed in the sedimentary deposits. On decomposition these remains yielded hydrocarbons which were stored in the rocks, sometimes evenly distributed, as throughout the bituminous Utica shale; at other times collected in caverns. The geological structure necessary for the preservation of oil and gas seems to be an anticlinal arch with an impervious layer above and a porous one below, or else a cavern in an impervious stratum. Some geologists hold that oil and gas are always the result of secondary distillation—that after the production of bituminous shales slow distillation takes place, and the products collect where the structure is suitable, or slowly escape. On this view oil should never be found in the rock in which the organic remains abound, but above it. For some fields, as the Ontario one, this is certainly not the case. Some have assumed that oil and gas are the more volatile parts of the immense mass of vegetation whose remains form our coal beds. The great oil and gas wells are, however, sunk in Silurian and Devonian strata, and consequently lie below the coal beds, which belong to the later Carboniferous period.

When a well is drilled into a petroleum pool, oil, gas, or salt water may be found. They are probably arranged in the porous sandstone in the order of their specific gravities, with gas at the top, water at the bottom, and oil between. Through long-continued distillation in a confined space, the gas is usually under great pressure. When the bore-hole reaches the deposit, the expanding gas either rushes out itself, or, if the bore tapped the cavern nearer the bottom, forces out the oil, or water, as the case may be. After exhaustion of the gaseous pressure pumping is resorted to. Before leaving a pumped-out well it is customary to "shoot" it. A charge of nitroglycerine is exploded in the bottom, by which new channels are opened and a fresh supply of oil often obtained.

Canadian Oil Fields.—In 1862 the first flowing well was struck at Oil Springs, Lambton county, Ontario. There was an immediate rush to the field. Dr. Alex. Winchell, in his "Sketches of Creation," describes the excitement and waste as follows "Though western Pennsylvania has produced numerous flowing wells of wonderful capacity, there is no quarter of the world where the production has attained such prodigious dimensions as in 1862 upon Oil Creek, in the township of Enniskillen, Ontario.

The first flowing well was struck there January 11, 1862, and before October not less than thirty-five wells had commenced to drain a storehouse which provident Nature had occupied untold thousands of years in filling for the uses—not the amusement—of man. There was no use for the oil at that time. The price had fallen to ten cents per barrel. The unsophisticated settlers of that wild and wooded region seemed inspired by an infatuation. Without an object, save the gratification of their curiosity at the onwonted sight of a combustible fluid pouring out of the bosom of the earth, they seemed to vie with each other in plying their hastily and rudely erected 'spring poles' to work the drill that was almost sure to burst, at the depth of a hundred feet, into a prison of petroleum. Some of these wells flowed three hundred and six hundred barrels per day. Others flowed a thousand, two thousand, and three thousand barrels per day; three flowed severally six thousand barrels per day. . . . Three years later that oil would have brought ten dollars per barrel in gold. Now its escape was the mere pastime of full-grown boys." Five million barrels were wasted in this way the first summer.

There are two distinct fields in Lambton county, separated by a synclinal fold. The Petrolia one extends west-north-west thirteen miles, and is about two in width. The Oil Springs field covers about two square miles. In both cases the oil is found in the Corniferous limestone—at Oil Springs at a depth of 370 feet; at Petrolia, 465 feet below the surface.

The following is the log of a well at Petrolia :

Surface	104 feet	} Drift.
Limestone ("Upper lime")...	40 "	} Hamilton.
Shale ("Upper soap").......	130 "	
Limestone ("Middle lime")..	15 "	
Shale ("Lower soap").......	43 "	
Limestone, hard white	68 "	
" soft	40 "	} Corniferous.
" grey.............	25 "	
Oil at a depth of	465 "	

About ten thousand wells are now in operation, yielding on the average about half a barrel a day. About four hundred wells are drilled annually to replace those exhausted. Pipe lines are laid through the district, and the companies receive oil from producers and store it until sold to the refiners.

A little south-west of Bothwell, Kent county, is a third field, which is likely to become a producing area. Small amounts of oil have been obtained in other parts of Ontario, notably Oxford, Essex, Perth and Welland counties and on Manitoulin Island; but no paying wells have been found. Recent discoveries on Pelee Island are promising. Oil oozes to the surface over a considerable area to the south of Gaspé Bay, Que. Several borings have been made, but the yield has been small. The prospect for productive oil wells is, however, a good one. In Nova Scotia and New Brunswick surface indications of oil have been found, but boring operations have resulted in entire failure.

In the valley of the Athabasca, in the North-West Territories, there is an immense deposit of tar sands. These sands are siliceous in character, fine-grained and cemented together by maltha, or inspissated petroleum. They belong to the Dakota formation, the lowest division of the Cretaceous, and lie unconformably on Devonian limestones. They outcrop over an area of one thousand square miles, and possibly extend beneath the surface as far as the Saskatchewan. In many places one-fifth of the sand, by bulk, is bitumen. It has been calculated by McConnell that there are six and a half cubic miles of bitumen in the Athabasca valley. It is the residue of a flow of petroleum from the underlying Devonian, unequalled elsewhere in the world. These tar sands will doubtless soon become of value as a source of bitumen.

Farther to the south there is a probability of finding oil which has not lost its volatile ingredients. South of Boiler Rapids the tar sand is overlaid by impervious shale, which in small anticlines doubtless has imprisoned some oil and gas. All through the Mackenzie River valley similar deposits of tar are found, and the same probabilities of extensive oil pools exist. In the South Kootenay Pass there are some indications of economic deposits being found in Cambrian strata.

Refining and Use.—The crude oil is distilled in large sheet-iron retorts. The easily vaporized gasoline and naphtha come off first and are condensed; then the kerosene, the wool oils, and lastly the lubricating oils follow; a carbonaceous mass is left behind. The coke is used as fuel; the other distillates are further separated and purified by redistillation and by chemicals. The Ontario oil contains a very large percentage of sulfur, and in the early days it was not known how to remove this. Canadian oil, as a result, had a disagreeable odor, and there is a prejudice against it to this day, though it is claimed that the best quality is now as good as any on the market.

The crude petroleum yielded the refiners in 1889:

Illuminating oils...................	38.7 per cent.
Benzine and naphtha	1.6 " "
Paraffin and other oils (including gas, paraffin, black and other lubricating oils and paraffin wax)...	25.3 " "
Waste (including coke, tar and heavy residuum).....................	34.4 " "
	100.0

Few raw materials yield as many products ministering to the comfort and happiness of men as does the rank-smelling crude petroleum. The benefits of cheap illuminating oil can hardly be overestimated. The lighter oils are used to mix the paints with which we adorn our homes, and the heavier vaseline is used to anoint our heads. Thick, black oils are used to lubricate car-axles and other heavy machinery, and white paraffin forms the basis of chewing gum. As solid paraffin, as liquid oil, as gaseous gasoline, petroleum affords us both heat and light. As naphtha and benzine, it is used as a solvent of fats.

Production.—The following tables show the magnitude of the oil industry:

PRODUCTION OF CANADIAN OIL REFINERIES.

Products.	1895.	
	Quantity.	Value.
Illuminating oils ....gallons	10,711,000	$1,217,000
Benzine and naphtha .... "	642,000	63,000
Paraffin oils .... "	1,016,000	140,000
Gas and fuel oils.............. "	6,095,000	219,000
Lubricating oils and tar .... "	1,699,000	76,000
Paraffin wax................pounds	1,840,000	83,000
Axle-grease .... "	......	8,000
		$1,806,000
Total crude oil used.......gallons	24,955,000	......

IMPORTS AND EXPORTS OF OIL AND ITS PRODUCTS.
1895.

	Imports.		Exports.
	Quantity.	Value.	Value.
Illuminating oils........gallons	6,471,000	$525,000	$3,000
Crude and lubricating oils "	1,107,000		
Paraffin wax ....pounds	164,000	12,000	....
Paraffin wax candles.... "	19,000	2,500	....

PRODUCTION OF PETROLEUM IN THE WORLD, 1894.

In Metric Tons of 2,204 lbs.

1.	United States	6,158,000
2.	Russia	4,873,000
3.	Austria	132,000
4.	Canada	116,000
5.	Roumania	75,000
6.	India (1893)	31,000
7.	Germany	17,000
8.	Japan	15,000
9.	Italy	3,000

—*Rothwell's " Min. Industry."*

LITERATURE.—Ontario : Geol. Sur. Reports, 1863, 1866, Q, S and S S, V. 1890-91; Min. Res. Ont., 1890. Gaspe : Geol. Sur. K, 1888. Kootenay : Geol. Sur. 1891, 9 A. Athabasca : Geol. Sur., 144 S, 1890-91. Bibliography : Rep. Q, 1890 ; Canadian Mining Manual, 1896. For complete description of the petroleum industry, see Crew, "Practical Treatise on Petroleum," 1887. For geology of petroleum, see Orton, An. Rep. U. S. Geol. Sur., 1889.

NATURAL GAS.

Burning springs have been known in many localities in North America from the earliest settlement, but with few exceptions, as at Fredonia, N.Y., no use was made of them. After the discovery of oil, large quantities of gas were frequently found in drilling for the former. For a number of years, however, even these bountiful supplies failed to attract attention. In 1879 gas was introduced into a Pittsburg factory, and from that time on its economic importance has been fully recognized and deposits of it eagerly sought. Few parts of North America are entirely destitute of reservoirs of gas, but the productive wells are almost entirely in New York, Pennsylvania, Ohio, Indiana and Ontario. Some gas fields are intimately associated with petroleum deposits, and the gas is doubtless of the same origin. In Ohio the Trenton limestone is the great reservoir, but in Ontario that formation is almost barren. It is in the Medina and Clinton divisions of the Upper Silurian that the Ontario gas is found. The Pennsylvania gas occurs in a still later formation—that of the Upper Devonian. A small amount of gas is found in the Cretaceous of the North-West.

Gas, like oil, has accumulated in porous rocks or under the arch of an anticline, overlaid by an impervious layer of shale or clay. It is the product of the distillation of plants and animals entombed in a sedimentary deposit. The distillation has gone on slowly for ages, the gas accumulating under pressure. On tapping the reservoir pressure is relieved and the gas escapes. Millions of cubic feet have been wasted, people not realizing that it was a store easily exhausted. This is well shown in the case of Pennsylvania, whose production has fallen from $18,000,000 in 1888 to $8,000,000 in 1891. Natural gas is a mixture of a number of gases, most of which are found in ordinary illuminating gases but in a different proportion. The following analyses from Sexton's "Fuel" will make this relation clear:

	Natural Gas.	ILLUMINATING GAS.	
		Coal Gas.	Water Gas.
Carbon dioxid and nitrogen	1.3	2.1	2.6
Marsh gas, CH_4	95.2	51.2	....
Heavy hydrocarbons $C_n H_{2n}$ ...	0.5	13.1	....
Carbon monoxid CO	1.0	7.8	20.2
Hydrogen	2.0	25.8	77.2

Canadian Localities.—Small quantities of gas from superficial deposits are found in many parts of the Dominion. In the North-West Territories some paying wells have been opened along the Canadian Pacific Railway, and on the Athabasca promising indications are found. The only localities of importance at present are in Ontario near the shore of Lake Erie. The Essex field extends east and west for a distance of twelve miles along the coast and for about two miles back. The wells are a little over 1,000 feet in depth, and yield from nothing up to 10,000,000 cubic feet a day. Two pipe lines carry the gas thirty miles to Windsor and Detroit.

The other district extends forty-five miles eastward from Cayuga nearly to the Niagara River. The gas is found in Medina sandstone at a depth of 700 to 850 feet, and issues from the wells under a pressure reaching in some cases to 500 pounds to the square inch. Pipe lines are laid through the district, and the wells are connected directly with Buffalo, where most of the gas is consumed. It is also used locally for burning lime and for lighting several towns and villages. Leamington, Ont., is said to have reduced its rate of taxation one-half by means of the revenue derived from supplying the village with gas. In 1895, 123 wells produced in Ontario 3,320,000 M. cubic feet of gas valued at $283,000.

ASPHALT.

Asphalt or bitumen is a mixture of various hydrocarbons, some of which are usually oxidized. It is a black or brown solid with a resinous lustre and bituminous odor, found as a superficial deposit in many parts of the world, but usually associated with bituminous rocks. Commercial asphalt is largely brought from a pitch lake on the island of Trinidad. Many varieties of asphalt have received distinct mineralogical names: of these albertite and maltha occur in economic quantities in Canada. All have been formed from petroleum by the vaporisation of the more volatile hydrocarbons.

The immense beds of maltha in Athabasca have been described under petroleum. Albertite is a pitch-like mineral found in the Lower Carboniferous of Kings and Albert counties, New Brunswick. At the Albert mine it occurred in an irregular fissure having a maximum thickness of seventeen feet. The veins are found in or near the Albert shales, a highly bituminous, calcareous clay rock with an abundance of fossil fish, and the mineral has apparently resulted from a distillation of this shale. Its composition, represented by 58 per cent. of volatile matter and 42 of fixed carbon, made it of great value for gas making, and 200,000 tons were shipped to the eastern United States for that purpose. The locality is now exhausted.

Anthraxolite is a name applied to a black combustible, coal-like substance found in Ontario and Quebec, which resembles anthracite in general characters. In composition it is essentially carbon, with from three to twenty-six per cent. of volatile matter. It never occurs in beds like coal, but in fissures in limestones, shales and sandstones. Dr. Sterry Hunt says, "It can scarcely be doubted that the coaly matters of the Quebec group have resulted from the slow alteration of liquid bitumen in the fissures of the strata." Some of the numerous occurrences may yield a few tons of fuel for local use. A vein at Sudbury is being exploited for this purpose.

Bituminous shales are often distilled for oil and gas. Works once existed at Collingwood and Whitby, Ont., for this purpose, but the discovery of petroleum destroyed the industry. Similar rocks were at one time distilled in Albert County, N.B., and in Pictou, N.S. The former yielded 63 gallons of oil and 7,500 feet of gas to the ton. When our petroleum deposits are exhausted these reservoirs of hydrocarbons may once more be of value. Similar rocks supply considerable oil in Scotland, competing successfully with American petroleum.

LITERATURE.—For description of the wells, production, etc., Geol. Sur. Reports Q 1890, S 1890, SS 1891, S 1892, S 1894 and Rep. Bur. of Mines, Ont., 1891. Bibliography, Geol. Sur. Q 1890. Origin—Geol. Sur. Rep.Q 1890 and Bur. Mines, 1891. Nat. gas in U.S., Ashburner, Trans. Am. Inst. Min. Eng. Vol. XIV., XV., XVI. Asphalt, Athabasca—Geol. Sur. 64 D 1890, 6 A 1894. Albertite, N.B.—Dawson, Acad. Geol ; Geol. Sur. 1876-7. Anthraxolite—Rep. Geol. Sur. 18 T 1888-9 ; Bur. Mines, Ont., 1896.

SECTION III.

ROCKS AND THEIR PRODUCTS.

CHAPTER XIV.

GRANITE AND SANDSTONE.

AMONG the materials which the mineral world furnishes for man's use, few are more important than those adapted for building. True, granite and clay and sand are so common to us Canadians that we hardly think of them as contributing to our mineral wealth. Nevertheless, one-quarter of our annual mineral production—that is, a little over $5,000,000 in value—is derived from rocks. A rock has already been defined as a variable mixture of minerals ranging in cohesion from loose débris to the most compact stone. Rocks are never the source of our useful metals, nor do they as a general thing yield us valuable chemical products. Their economic importance lies, for the most part, in their structural adaptability. No other material approaches them in strength or durability. The extent of our forests and the consequent cheapness of timber have caused us to neglect our granites and limestones. As lumber increases in price and as the need for more indestructible buildings grows, there will doubtless be a greater employment of stone. True, many farm-houses are built of boulders, and some of our towns are quite largely erected from limestone quarried in the neighborhood. In both cases cheapness has been the only desideratum, and durability and beauty have been neglected.

Building Stones.—That a rock be useful as a building stone it is necessary that it should be strong and durable. It is also desirable that it be easily quarried and dressed, and that it have beauty of color and texture. Strength and durability depend on several considerations. The finer the structure and the more compactly the grains are consolidated the greater the strength. The kind and amount of cementing material exerts a great influence on both strength and durability. A cement filling all the interstices of a rock will evidently make a stronger stone than one in which the grains are merely held together by their adjacent faces. A siliceous cement is stronger than a calcareous one—a ferruginous than an argillaceous. Again, a porous rock is capable of absorbing considerable water, and in our cold climate this is a deleterious property. As freezing water expands with enormous power, the outer parts of the stone are slowly forced off, and ultimately the whole crumbles. According to Merrill a rock which absorbs 10 per cent. of its weight of water in twenty-four hours should usually be discarded. Some good sandstones approach this amount; granites average perhaps one-twentieth as much.

Fineness of grain and uniformity of size are conducive to durability. In a granite, for instance, under the influence of the sun's heat all the grains expand.

And since the rate of expansion is different for each of the ingredients, mica, felspar and quartz, a strain is put on the cementing material. Alternate expansion and contraction ultimately results in disintegration. "Dr. Livingstone found in Africa that surfaces of rock which during the day were heated up to 137° F. cooled so rapidly by radiation at night that, unable to sustain the strain of contraction, they split and threw off sharp angular fragments from a few ounces to 100 or 200 pounds in weight." In burning buildings the heat is still greater, and the sudden cooling produced by dashes of cold water tests a stone severely. Granite, of all the rocks, is the least fire-proof. Marble and limestone are the least affected where the heat is not sufficient to cause decomposition and where water is absent. With greater heat sandstone is most resistant.

Another cause of decay is the presence of injurious accessory minerals. Pyrite is the most common and the most injurious. It slowly unites with oxygen to form the various oxids and hydroxids known as rust. In some cases only the beauty of the stone is marred; in others its strength is weakened. Ferrous carbonate and small seams of clay are other deleterious minerals.

The facility with which a rock may be worked depends on the hardness of its constituents and on the presence of joints, beds or other natural fractures. A granite is harder to work than a limestone because of the hardness of the quartz and felspar of the former. For a similar reason, also, a siliceous sandstone is more costly to market than an argillaceous one. A rock which cleaves regularly in any direction can be more cheaply produced than one with an irregular fracture.

In the selection of a building stone for important structures durability is of prime importance. The most reliable information can be got by observing the effect on old structures. Failing these, an examination of the natural outcrop of the rock will yield information concerning its weather-resisting power. "If in these exposures the edges and angles of the stone remain sharp—if its surface shows no sign of flaking or crumbling, no cracks nor holes where pyrites or clay has lurked, nor dark stains from the change of iron compounds—it may be relied upon for structures if proper care is used to reject suspicious blocks." Much also may be gathered from a microscopic examination. Of secondary importance is the strength, though this is the property which is most usually tested. Any compact stone has many times the strength usually required. Imperviousness to water would be a more desirable test. For piers of bridges, foundations and other rough purposes, faults of color, coarseness of texture or irregularity of fracture are of no account, and proximity and consequent cheapness will be the condition sought.

The Crystalline Rocks.—Immense areas of granite and allied rocks are found in Canada—a quantity sufficient to supply all the world with building stone. The commercial term granite includes not only the true granite of the geologist but a number of related rocks. Syenite has the general appearance of a granite, but is without the quartz of the latter. Both have orthoclase felspar and either mica or hornblende. Gneiss has the same minerals but is schistose in structure. All three are quarried for building purposes, and the granite and syenite for monumental stones. They are widely distributed through the whole Dominion, the region of the plains excepted. Granite is expensive to work, and has not yet been used to any extent in Canada as a building stone. It seems, however, quite unnecessary for us to import granite from Scotland for monuments when quite as good stone surrounds us on every side. Quarries have been opened in British Columbia, at Kingston and Gananoque, Ont., in Stanstead, Que., in New Brunswick and in Nova Scotia, from which about 13,000 tons are annually raised, valued at $70,000. These granite rocks, as well as the more basic igneous rocks, diorite, anorthosite, etc., are also used as paving stones.

Sand and Sandstone.—The crystalline rocks slowly disintegrate through the action of heat, moisture and frost, and the streams carry off the products to deposit them ultimately in some lake or ocean. The particles of quartz are much the most enduring. Felspar, mica and hornblende are not only separated from each other by the weathering of the rock, but are also decomposed. All three yield clay and some free silica, besides other minerals. The quartz, though rounded on the edges through long-continued rubbing, remains pure silica to the last. Thus it is that most rocks, when reduced to fine grains, yield a sand which is largely silica. Pure silica is white, and the light yellow color of many sands is due to stains of iron oxid or to a mixture of black grains of the magnetic oxid of iron. Small amounts of undecomposed mica or felspar may also be found. In a limestone region the sands may be calcareous. Clay also may be mixed with the sand.

Sands of all kinds are widely distributed over our country, and are in all cases a superficial deposit. Only on rocky hills, swept bare by glacial action, are they lacking. Sandstones are but consolidated sands. They have been formed in ancient seas by the pressure of overlying material, and have since been raised above the water. A cement of iron oxid, silica, clay or limestone holds the grains together, and gives a distinctive character to the rock. Some sandstones are almost pure silica; others through the presence of clay merge into shales; others again shade gradually into limestones. In some cases these sandstones were subjected to heat as well as pressure, and all the materials in them were recrystallized. Pure sand, metamorphosed in this way, became the solid white quartzite so common in our Huronian districts. A sand with mica became a mica schist; one with felspar and mica became a gneiss, and so the cycle was completed from igneous rock back to igneous.

Sandstones are usually bedded, the planes of stratification representing intervals in the deposit of sand on the ocean floor. The deposit of one period became somewhat consolidated before the next supply of material was brought down. The beds are sometimes but a fraction of an inch in thickness, at others several feet. The thicker beds which split readily in any direction are known as freestone.

In the very dawn of geological history sands were being deposited in Canada as they are to-day. Consolidated and metamorphosed they form the quartzite of the Huronian. Above them lie sandstones of Cambrian age. Silurian, Devonian, Carboniferous, Triassic, Cretaceous and Miocene times contributed their quota of sandy sediments. So through the whole Dominion sandstones are abundant and cheap. They are used extensively for building; also as flagstones, furnace linings, grindstones and whetstones. As powdered stone or as the natural sand, quartz is also used for mortar, glass, moulding and polishing.

Building Stone.—In the Maritime Provinces there are considerable areas of good freestone in the Lower Carboniferous rocks. The stone is soft enough to be readily cut when first quarried, but hardens on exposure. Red, yellow, light grey and beautiful olive-green beds are found. The stone is not only used domestically but also exported. The chief quarries are at Dorchester, Hopewell, and neighboring localities in Westmoreland and Albert counties, New Brunswick. Amherst, Wallace and Pictou in Nova Scotia also produce good stone, some of which is exported. The magnificent court-house of Toronto, Ontario, is constructed of New Brunswick stone.

In Quebec a sandstone of the Potsdam or Upper Cambrian period affords an excellent building stone. It is almost white in color and very hard and durable.

It is quarried, among other places, at St. Scholastique and at Hemmingford, and used in Montreal. It has also been used successfully at St. Maurice as a furnace lining. Near Quebec and Levis the Sillery sandstone is quarried and used quite extensively in both cities. It is usually a green or greyish-green rock, though on the coast below L'Islet there are beds of a purplish-red color. The rock does not weather uniformly, nor is it as durable as the Potsdam stone. Some Silurian sandstones have been quarried in Gaspé for railway work.

The Potsdam sandstone of Quebec occurs on the south of the Ottawa River in Ontario, and here, also, has been extensively used. Considerable was quarried in Nepean township for the national Parliament Buildings at Ottawa. Farther to the west a band of Medina sandstone outcrops along the Niagara escarpment, which stretches from Queenston Heights past Hamilton to Cabot's Head. It is quarried at a number of places, principally along the Credit River. The stone occurs in both white and red beds, the latter being the more valuable. It is very extensively used in western Ontario—the Parliament Buildings at Toronto being a good example of the appearance of the red variety. A similar red stone of Cambrian age occurs in the Nipigon formation on the northwest of Lake Superior. It has been shipped from Verte Island to Chicago and other lake cities.

In British Columbia freestone of Cretaceous age may be quarried at many points along the coast. Some excellent material for building has been obtained near Nanaimo. A white freestone of the same age is quarried at Calgary, Alberta.

Other Uses.—Flagstones have been obtained at most of the localities just described, and at many others. Material suitable for grindstones has been quarried at Seaman's Cove and other points in Nova Scotia, and in Albert and Westmoreland counties, New Brunswick. Some grindstones and coarse whetstones are made from the Medina in Nottawasaga, Ontario, and the Cretaceous of Nanaimo, British Columbia, is used for the same purpose. The total annual production of grindstones is valued at about $40,000, of which one-half is exported, chiefly from Nova Scotia. The imports about equal the exports.

Sand for mortar-making should consist of sharp angular grains of quartz of somewhat coarse texture. When an impure mixture of sand and clay is used the mortar frequently crumbles. Good material is widely distributed in the superficial deposits.

Sand for moulding is not at all plentiful. It is an "intimate mixture of quartz sand with just sufficient proportions of clay and ochre to enable it to retain the form given by the pattern." A good moulding sand contains about 92 per cent. of fine quartz sand, 6 per cent. of clay, and 2 per cent. of iron oxid. For fine castings, artificial mixtures are often prepared. Suitable sand is found at several places in Ontario and Nova Scotia. From Windsor, N.S., a small amount is annually exported.

Ordinary glass is made from quartz sand, sodium carbonate and lime. Except for the coarser varieties of glass, a fine, angular white sand is needed, free from all impurities, especially iron. Ordinary bottles have a green tint due to the iron of the sand. Many pure sands are found in the Dominion, and several sandstones could be crushed and used. The Potsdam sandstone was at one time used at Vaudreuil.

Sand is further used as an abrasive in sawing and polishing sandstone and marble. Tripolite, or infusorial earth, is also used as a polishing material under the name of "silex, electro-silicon," etc. It consists of the microscopic siliceous shells of diatoms and other minute water plants. Though each individual was so small, beds thirty feet thick have been formed extending over considerable areas. Many deposits are known in Canada, from which over 600 tons valued at $10,000 were taken in 1896. Tripolite was at one time used as an absorbent of nitroglycerine, and is now employed in the manufacture of water filters.

LITERATURE.—Merrill, "Stones for Building and Decoration," gives a full account of the properties of building stones and of methods of working. For details of Canadian quarries, see Dawson, Acadian Geology; Geol. Can., 1863; Min. Res. Ont., 1890; Bur. Mines, Ont., 1891; Rep. R., Geol. Sur., 1887; Rep. S., 1894. For localities of various sands, tripolite, etc., see Cat. Sec. 1 of the Museum of the Geol. Sur.

CHAPTER XV.

CLAY AND SLATE.

AMONG mineral materials few are more important than common clay, although it is so widely distributed that we often forget our great dependence upon it It ministers to our wants in numerous and in very diverse ways, the products often bearing no apparent relationship to one another. Sun-dried bricks and porcelain dishes are entirely different in appearance. Clear, transparent china bears little resemblance to drain-tile, and yet all four are essentially the one thing—clay. The manufacture of rude pottery was one of the first arts practised in the dawn of civilization, and ceramics has advanced step by step with man's development. The value of our clay output to-day is only exceeded by that of our coal.

Origin and Composition.—Clay is not an original mineral, but the product of decay—the result of the passage from an unstable compound to a stable one. The felspars which are found abundantly in igneous rocks are easily attacked by water and carbonic acid. They are all silicates of aluminum, with potassium, sodium or calcium. The potassium felspar, orthoclase, is the most abundant. This mineral, and the others as well, lose their alkaline constituents together with some of their silica, and take up water. The alkali goes off in solution, and the silica and hydrous silicate of aluminum are left. This last, when pure, is known as kaolin. Its composition is represented by $H_2Al_2(SiO_4)_2 + H_2O$, or silica 47, alumina 39, water 14 per cent. Usually there is mixed with it some quartz and mica of the rock, some undecomposed felspar particles, and some oxid of iron, calcium carbonate and alkalies, the accessory products of decomposition. Commercial clay may be the pure kaolin or any of the numerous mixtures possible. In some of the best clays kaolin is much the largest ingredient; in others, considerably less than half. It is the essential constituent—the other minerals are but accessories, and often injurious ones. Quartz, in the form of fine sand intimately mixed with the kaolin, is the most common impurity. By itself in a clay, silica is chemically inert but acts physically, checking shrinkage and cracking when the kaolin is highly heated. When potash, soda or lime are present the silica unites chemically with them at high temperatures, forming fusible compounds which give strength and hardness to the pottery. Some of these alkalies are nearly always present—potash most commonly. Magnesia often replaces lime. Iron, either as an oxid, carbonate or sulfid, is the most undesirable impurity and is nearly always present. Sometimes it does not constitute more than one-fifth of 1 per cent.; more frequently it makes two to ten per cent. or more of the clay.

Clays resulting from the decomposition of felspars in place are classed as residual clays. They nearly all contain quartz, which is easily removed by washing. They often exist as a crumbling rock resembling granite. The chief characteristic of this residual, or rock, kaolin is its non-plasticity. These residual clays are, of course, subject to the erosive and transporting action of water, and immense beds of sedimentary clays have been deposited in quiet waters since the beginning of geological history. They are always more or less impure and are generally highly plastic, a property probably due to the rubbing the particles have undergone. They form the chief basis of the world's clay industries.

Those deposited in Palæozoic times have, for the most part, been consolidated into shales, and many of them have even been metamorphosed into slates. The latter have ceased to have a value in ceramics, but the former are very widely used, after being ground and allowed to weather. The Carboniferous period furnishes a valuable refractory clay. Cretaceous, Tertiary and Quaternary clays are extensively used in America. In the last era ice, not water, was instrumental in producing deposits of clay which are not residual. Boulder clay, as it is called from the angular stones it contains, resembles sedimentary clay in its composition and properties, but lacks stratification.

Uses.—"The chief function of clay in the fictile arts is its partial fusion upon firing, and upon this and the skill of the artisan who fires the kiln depends the product, which is wonderfully varied by the mixtures of fluxes and tempering material. Plasticity is desirable for the handling of the unfired material. Nearly all unconsolidated or powdered rock material may be made to adhere by water and other ingredients than clay, so that it can be shaped for burning, but plastic clay is the cheapest material used for this purpose in all clay-burning." (Hill, Min. Res. U.S., 1891.) Clay is used in the manufacture of a number of domestic utensils, as porcelain, China and earthen ware. As a structural material it finds employment as brick, terra cotta, roofing tile, draining tile, door knobs and sewer pipe. In the industrial arts it is used as a lining for kilns, furnaces and retorts; for crucibles, for moulding-material, as a base for pigments, for filling paper, and even as a food adulterant.

Commercially clay may be divided into four classes, depending partly on composition and partly on use. Chemical composition is not the sole guide in determining the value of a clay, for those almost identical in composition often yield different products on firing.

1. China clays are nearly pure kaolin and non-plastic. They are nearly always ground and washed before use, but should be free from iron and lime. Mixed with felspar and silica they are used to make China ware. Cornwall, Limoges in France, and Dresden in Germany have important deposits of these rare clays.

2. Plastic, ball or pottery clays are the essential material of bricks, pottery and stone ware. The

purer ones are China clays in composition, but will not yield the same products on firing. These clays are used in the production of earthen ware, etc., and to give plasticity to China clays in the manufacture of China ware. Deposits near St. John, Que., are used extensively in the production of porcelain.

3. Brick clays include those suited not only for the manufacture of bricks, but also of drain tile and the cruder kinds of stone ware. They are most widely distributed of all, and, probably, are most important economically. Ideal brick clay consists of a mixture of fine sand and pure plastic clay, the proportions of which may vary very widely. A good clay consists of three-fifths silica, one-fifth alumina, and the remainder of iron, lime, soda, potash, magnesia and water.

Iron is present in most brick clays and is the basis of color. Red bricks are produced from white clay by the oxidation of the iron from the ferrous to the ferric compound. Still, as is well known, the color may be modified by differences in the temperature of the kiln. White bricks are often supposed to be due to the lack of iron in the clay, but the correct reason seems to be that these clays contain lime or magnesia, which unites with the iron and with silica to form a colorless silicate.

Vitrified bricks are being introduced into Canada as a paving material. They offer all the advantages of asphalt and are considerably cheaper. A vitrified brick may be described as a piece of clay heated to incipient fusion, so that all the particles have been fritted together and the pores have become closed. Its excellence is measured by the degree with which water is excluded. To be suitable for this purpose a clay must agglutinate or vitrify some distance below its point of fusion, otherwise in the firing much of the product will be destroyed by melting. Several companies are making these bricks near Toronto.

All of these clays are widely distributed through the Dominion. The shales of the Hudson River and Medina epochs are used in Ontario to make a very fine pressed brick. Sewer pipe, drain tile and póttery are made at so many points that it is useless to enumerate.

4. The refractory, or fire-clays, form the last division. Alkaline fluxes are here present in very small quantities. Pure kaolins are desirable as the base of the mixture, which is usually made artificially. The Cretaceous clays of New Jersey and the Carboniferous under-clays are often suitable. A number of fire-clays of fair value occur in the rocks of the latter period in Nova Scotia.

The production of these materials in 1895 was valued, as follows: Building brick, $1,670,000; terra cotta, $195,100; sewer pipe, etc., $257,000; pottery, $151,600; fire-clay, $3,500; a total of $2,277,200. In the same year the imports amounted to $593,300, most of which was for earthen ware.

Slate.—When a bed of clay or shale is subjected to great pressure and heat its physical characters are changed. The laminæ become smooth and hard, and microscopic crystals are often developed throughout the fragmental material. Minute flakes of mica are usually present, their flat surfaces being parallel to the face of the lamina. The well-developed cleavage is rarely parallel to the original plane of bedding, but is at right angles to the direction from which the pressure came. Under this pressure the component grains of the original sediment rearranged themselves with their longest axes at right angles to the direction of force, and so made new planes of cleavage.

A number of varieties of clay slate are recognized. Roofing slate includes the finest-grained, compact kinds used for roofing houses, for mantels and table-tops, for slates and pencils, etc. Whet-slate or honestone is a hard, fine-grained siliceous rock. Phyllites embrace the thoroughly metamorphosed shales characterized by the development of much mica and the recrystallization of the materials.

These slates are found in the majority of the geological horizons, but the Huronian, Cambrian, Silurian and Devonian formations contain them most frequently. Good roofing slates are found in Canada in the Cambrian rocks, east of the St. Lawrence. Quarries are worked at New Rockland, Shipton, and near Richmond, all in Richmond county, Quebec. A number of other quarries have been opened in neighboring counties, but the demand does not justify their operation. The usual color is dark or bluish-grey, but green, red and purple ones are found. The best class cleave readily, are "free from pyrites, impervious to water, and equal in every respect to the celebrated Welsh slates." Roofing slates, slabs and school slates are produced in this district. The product in 1895 was valued at $59,000, about one-half that of 1889. The imports in 1895 amounted to $19,000, also about half of the corresponding figures for 1889. A small amount is annually exported.

LITERATURE.—"Clay Materials," by Hill, in Min. Resources of U.S., 1891, contains a good description of the kinds and uses of clay. See also Geol. Can., 1863. "Brick Clays of Que.," Rep. Geol. Sur., IV. 188 K; "Brick Clays of Ont.," Bur. of Mines Rep., 1891, 1893, 1895. The report of 1893 contains a chapter on vitrified brick. "Fire Clay of N.S.," Rep. Geol. Sur., V. 1890, 190 P; "Slate of Que." Rep. Geol. Sur., IV. 1888 K.

CHAPTER XVI.

LIMESTONE.

Origin and Occurrence.—Limestone is one of the most widely distributed rocks occurring in all the sedimentary formations from the Cambrian down to recent times. It is found even in Archæan areas as great bands of crystalline material which are metamorphosed sediments. Geographically its distribution is as wide as it is geologically, and every province but Prince Edward Island has its own supplies. The only large areas of the Dominion destitute of it are some of the districts covered by the igneous Archæan rocks.

It has always been deposited as a sediment, sometimes as a chemical precipitate, much more frequently as a bed composed of the fragments of the shells and skeletons of lime-secreting animals. As is well known, gravel and sand derived from the land are deposited near the shore and the lighter mud carried farther out. Beyond this, where sediments from the land were rarely brought, the bottom of the old ocean beds was slowly built up by the calcareous remains of dead molluscs, crinoids, corals and other organisms. The process can be watched to-day on the coast of

Florida, and time and the pressure of superincumbent beds are alone needed to transform the loose shell deposits of that peninsula into solid limestone. Consolidation and recrystallization are promoted by the easy solution and precipitation of calcium carbonate in waters carrying carbonic acid.

Often these deposits were made when mud or sand was being laid down, so that beds of limestone and shale or of limestone and sandstone are now found to alternate with one another, and even to pass by gradual changes from one into the other. A pure limestone consists of calcium and carbonic acid, that is, it is the mineral calcite ($CaCO_3$). Frequently the calcite is replaced by dolomite, an isomorphous mixture of calcium and magnesium carbonates. Silica, clay, oxids of iron and bituminous matter are often present as impurities. The color is commonly a dull white to a blue-grey, but may be brown or black. Few rocks vary more in texture than limestone. It may be a hard compact rock with a choncoidal fracture; it may consist of crystalline grains resembling loaf sugar in texture and color; it may be an earthy, friable deposit, or a compact rock resembling a close-grained sandstone. In all cases it is easily scratched with a knife, and gives a vigorous effervescence when treated with hydrochloric acid.

Uses.—Limestone is probably the most valuable of all our structural materials, for not only is it an excellent building stone itself, but it also affords the most useful cement for holding all other building materials together. It is employed not only in the

THE MINERAL WEALTH OF CANADA. 181

farm-house but in the city cathedral; it is used not only for the outer walls but also as marble for the decoration of the interior. It is used for bridges and culverts in railway construction, and for the concrete foundations of city pavements. As a flux in the smelting of iron it finds a large employment, over 30,000 tons being annually used in Canada alone, where the iron industry is not a large one. Some fine varieties are used as lithographic stones. Marl, an amorphous mixture of calcium carbonate, clay and sand is a valuable fertilizer. (See Chapter XVII.) Chalk, a soft earthy variety of limestone not found in Canada, is used by carpenters and others for marking; perfectly purified and mixed with vegetable coloring matters, it forms pastil colors. Whiting is a purified chalk used as a pigment and as a polishing material.

The desirable qualities in a limestone to be used for structural purposes have already been pointed out (Chapter XIV.), and it is only necessary to indicate here some of the important localities where stone occurs. Limestone is so widely distributed throughout the Palæozoic areas of southern Ontario and Quebec, and of Nova Scotia and New Brunswick, that it is useless to attempt an enumeration of the places where it is quarried. The lowest horizon to furnish valuable stone is the Chazy, which is extensively quarried at St. Dominique, Phillipsburg and Montreal Island. The Trenton limestones, occurring in the neighborhood of Montreal, also furnish that city with excellent building stone. In Ontario, the Niagara formation is worked at a number of places along the escarpment which enters the Province at Queenston and passes by Hamilton and Owen Sound to Manitoulin Island and into Michigan. Stone from Queenston, Thorold, Beamsville and Grimsby has been extensively used in the Welland canal, the St. Clair tunnel, and railway construction throughout the Province. The Corniferous also gives a valuable stone where exposed. Quarries near Amherstburg furnished material for the Sault Ste. Marie canal. In Nova Scotia and New Brunswick Carboniferous limestone of excellent quality is widely spread, and is quarried in a number of places.

Marble.—The term marble is properly applied to a crystalline aggregate of calcite grains of uniform size, and each of which is composed of twin crystals with their own cleavage lines. It has been produced by the recrystallization of ordinary sedimentary limestone *in situ*, occasioned by the heat of eruptive rocks and the pressure of overlying masses. Typical marble is white, but it may be yellow, green, blue, black, banded or mottled. Sometimes it is very fine-grained, as in the best statuary marbles; again it may be so coarse as not to take a good polish, and so be useless for ornamental purposes. Mica, garnet, tremolite and many other species of silicates are frequently found in it, a result of the recrystallization of sand and clay impurities in the original limestone.

Commercially, the term marble is applied to any limestone, crystalline or non-crystalline, which is susceptible of a polish, and is suited in texture and color for ornamental work. It is even made to include serpentine, when this magnesium silicate is found in masses suitable for decoration. On the contrary, impure marbles and those of too coarse grain to be of value for decorative work are classed as limestones, and used for structural purposes.

True marbles are found in regions of metamorphism, particularly in the Laurentian areas in Canada. From the Georgian Bay east to the Ottawa valley are scores of bands of crystalline limestone interbedded, with gneiss and other schists. These have been worked to a small extent at a number of places, as at Madoc, Bridgewater, Renfrew and Arnprior in Ontario. Across the Ottawa it is found in Hull, Grenville and other places. A very fine marble of similar age is quarried at West Bay, Cape Breton. At Echo Lake, near the St. Mary River, Ontario, a close-grained limestone of Huronian age has been worked to some extent. It is composed of thin, alternate bands of grey and colored stone, and takes an excellent polish. In the metamorphic rocks of the Eastern Townships marble is quarried for local use at several places. At Dudswell a rock of Silurian age is entirely composed of organic remains, principally corals, which when polished presents a beautifully marked surface. The Eozoon limestone, which consists of an intimate and irregular mixture of white calcite and green serpentine, gives a handsome effect when polished. It is found in the Laurentian rocks in Grenville and Templeton, Que, and is supposed by some to be the remains of the earliest known animal. Serpentine, which occurs in large masses in the Eastern Townships, is used for interior decoration under the name of verde antique marble. At Texada Island, B.C., a grey, white and mottled stone is quarried and used for monumental and decorative work.

With such large and varied deposits of marble it is strange that we depend so much on other countries for our supplies. For the past ten years our production has averaged only 300 tons, valued at less than $5,000, while the imports amount to over $100,000 a year.

Lithographic Stone.—Limestones of fine even grain, entirely free from crystals of calcite, are extensively used in the duplication of maps and drawings. It is almost impossible to define the characteristics of a good lithographic stone, for in both chemical composition and physical structure the few suitable limestones are exactly imitated by hundreds of useless ones. The stone of Solenhofen, Bavaria, which is used the world over, is an even-grained, compact limestone, with less than 6 per cent. of clay and other silicates. It is buff or drab in color by reason of a small amount of organic matter, which is, perhaps, the most valuable constituent. Suitable material has only been found in Bavaria, Silesia, England, France and Canada. In the last-named it occurs as a number of beds six to twelve inches thick in the Trenton limestone in the township of Marmora, Ontario. In composition and physical characters it closely resembles the Bavarian stone, which is of Jurassic age. Several quarries have been opened, and trial shipments have shown some of the stone to be of excellent quality.

Mortar and Cement.—Among the mineral cements there are none which approach in importance those which consist of lime or some of its compounds. Ordinary mortar is made from quick-lime and sharp clean sand, its cementing qualities depending chiefly on the formation of calcium carbonate by the absorption of carbonic acid from the atmosphere. At the same time calcium silicate, which forms very slowly, considerably strengthens the cement after a number of years. Both ordinary limestone and dolomite are converted into lime by heating in kilns until the carbonic acid has been expelled. The first yields "hot" limes, the latter "cool" limes, so called from the relative amounts of heat developed in slacking. Both form good mortars, although the magnesium limes slack less rapidly and set more slowly. Both varieties are extensively made in Canada, particularly where other limestone industries are established. Every province except Prince Edward Island has its own supplies, the total product being valued at $700,000 in 1895.

Ordinary lime like that just described, which is made from nearly pure material, will not harden if immersed in water, but if made from a rock containing considerable clay it has this valuable property. Such a lime is properly called a cement, and it may be a natural or a Portland one, according as it is made from natural rock or an artificial mixture. A hydraulic limestone consists, then, of calcium or magnesium carbonate mixed with fifteen to thirty-five per cent. of clay and a little alkali. Such a rock on being strongly heated forms a double silicate of calcium and aluminum, a compound capable of uniting with water to form a hard, crystalline compound, even when immersed.

Hydraulic limestones are widely distributed, and are converted into natural cement at a number of places. The rock is burned in kilns like ordinary lime, and then, since it does not slack at all with water, or very slowly, it is ground to a fine powder. The product often lacks uniformity, for the chemical composition of the beds of a quarry vary greatly. For this reason artificial cements are often preferred. The original Portland cement was made by grinding together a mixture of clay and chalk of definite composition and then calcining and regrinding. Artificial cements are now made at a number of points in Canada, as at Napanee and near Owen Sound, Ont. The production of cement in 1895 was 128,000 barrels, most of it coming from Ontario, and nearly half of it being classed as Portland. The total value was $174,000. In the same year the imports of all kinds of cement amounted to $252,000.

LITERATURE.—Marble: Min. Resources of Ont., 1890; Rep. Geol. Sur., IV. 1888 K. Lithographic stone: Rep. Bur. of Mines, Ont., 1892, 1893. Cement: Bur. of Mines, Ont., 1891; Gillmore, "Limes, Hydraulic Cements and Mortars."

CHAPTER XVII.

SOILS AND MINERAL FERTILIZERS.

AMONG the varied resources of Canada none is of greater importance than her fertile soil, the direct support of more than half of the population. Nor is there need of any excuse for introducing here a short chapter on soils, for the connection between geology and agriculture is of the closest character, though it is unfortunately too seldom recognized. The origin and distribution of soils; the cause of their fertility; the source and proper use of minerals to restore the necessary losses incurred in cropping, are questions of a geological character of the first importance to the progressive farmer. To the student, also, the transformation from the hard and barren rock to the loose and fertile soil is of exceeding interest. The uses of rocks in their original, living state are not to be compared with their value to man after old age has overtaken them and death and decay have reduced them to dust. This finely divided rock material, constituting the superficial portion of the earth's crust, is known as soil. It is composed chiefly of very variable mixtures of clay and sand, with considerable proportions of vegetable matter and iron oxid.

Origin of Soil.—As soon as a sedimentary rock appears above the water, or an igneous rock is extruded from the crust, meteoric forces begin to transform it. Wind and water, heat and cold, plants and animals, oxygen and carbonic acid, all unite to disintegrate and dissolve the solid rock, and even to transport much of it to other localities. Water, oxygen and carbonic acid are the chief agents involved in producing chemical change. The ferrous and the manganous compounds, so frequently constituents of igneous rocks, easily take up oxygen to form the more stable peroxids. Sulfids of the metals become soluble sulfates, and these may even lose their sulfuric acid and be precipitated as hydrates, as in the transformation of iron pyrite into limonite. Rain water always contains some carbonic acid, and as it percolates through decaying vegetable matter it soon becomes charged with this powerful solvent. The silicates of lime, soda, potash and iron, so abundant in the crystalline rocks, are easily attacked by this water, carbonates of the bases being formed and silica set free. The crystals of felspar lose their lustre and color, first becoming dull and earthy on the outside, and finally being converted into a soft, pulverulent clay. The rapidity and completeness of the process vary greatly, but usually all of the alkalies and much of the silica are removed. Water charged with carbonic acid is also a good solvent of ordinary limestone, calcium carbonate being carried off and the impurities left behind.

Solution is greatly aided by physical disintegration.

Mosses insert their tiny rootlets and open the way for other agents. Larger plants, by the power of their growing roots, wedge off pieces of rock, and so promote chemical solution. The unequal expansion of different minerals when subjected to the heat of the sun has a disintegrating effect. Most powerful of all these influences is that exerted by freezing water. All rocks absorb a little moisture, and those that are porous or fissured are particularly susceptible to the destructive effects of frost. The angular blocks on every mountain slope attest the power of this agent.

Abrasion also promotes disintegration and consequently decay. Running water rolls the broken rocks over and over, wearing off the angles and gradually reducing them to sand and gravel. The shore ice of rivers, lakes and seas often surrounds large stones, and driven by the wind or current, abrades both them and the shore. Still more potent was the ice-sheet which at one time covered Canada, as it does Greenland to-day. This mantle of ice moved slowly downward from the Laurentian heights, carrying in it and under it great blocks of granite and other igneous rocks which, pressed against the underlying ones, were slowly ground to pieces. Abrasion, disintegration and chemical change have thus transformed the barren rocks into fertile soil.

Classification.—In accordance with their origin two classes of soils are recognized, sedentary soils and transported soils. The first class are comparatively rare in North America north of the thirty-ninth parallel of latitude, the point to which the ice-sheet extended. South of this line they are the prevailing class, except in the river valleys. Soils derived from the disintegration of sandstone are of course very sandy, containing only the small amount of clay present in the original rock. Shales and soft slates weather to clay soils undesirably heavy and compact, except where the shale contained considerable sand. The disintegration of a limestone is usually accompanied by solution, so that the resulting soil is largely composed of the original impurities, chiefly clay and iron. Indeed, a calcareous shale will, on weathering to a clay, retain much of the lime, while a soil resulting from the disintegration of a limestone may be nearly devoid of calcareous material. Sedentary soils formed from granitic rocks are usually thin and poor. When decomposition is very rapid, the felspars and micas yield a clay retaining some of the alkaline and calcareous ingredients of the original rock, and this mixed with the abundant silica furnishes a fair soil. All of these sedentary soils gradually merge by coarser materials into the rocks on which they rest.

Transported soils embrace those which have been formed through the agency of water or glacial ice, and which bear no relationship to the rocks beneath them. In Canada, those due to glacial action are by far the most extensive and among the most fertile. These soils have been spread over the country often to a depth of several hundred feet, obliterating frequently the old drainage systems and giving a new contour to the surface. They consist of clay and sand and gravel, derived often from very different sources and intimately mixed. The product of abrasion and not of decay, they contain all the elements of fertility found in the original rocks. Since their deposition the surface has of course been subject to the ordinary meteoric influences, and some of the soluble salts have been carried away. The subsoils, which have been subjected in a less degree to atmospheric agencies, are naturally richer in a number of ingredients necessary for plant growth. Proper tillage tends to restore to the surface what is being continually lost through the growth of crops and the solvent action of rain. Man accomplishes this by deep ploughing, and he is helped not a little by the action of worms and other burrowing animals.

Besides the "drift," there is another division of transported soils known as alluvium. This is water-carried material which may have been deposited in the flood plain of a river, in the basin of a lake since drained, or in the marshy inlet of a sea at high tide. These alluvial soils are frequently very fertile, containing as they do much of the best material borne from the higher lands. The fine silt brought down by the Nile has transformed its desert flood plain into rich agricultural land. The marsh lands of Nova Scotia and New Brunswick, among the most fertile soils of the Dominion, are due to deposits of silt made at high tide. Fifty thousand acres have been reclaimed by dikes around Chignecto Bay alone.

Soils are also classified according to composition.

They may be clayey, sandy, peaty or calcareous as one or other of these constituents predominates.

Fertility.—The fertility of a soil depends on its chemical composition and on its physical texture. The useful physical characters are (1) sufficient looseness to afford easy penetrability to roots, to moisture, to air and to fertilizers; (2) sufficient retentiveness to prevent a rapid loss of water and fertilizing material. These properties depend on the relative proportions of sand, clay and humus which constitute the soil. Too much sand makes a light soil easy of cultivation and readily dried, but not retentive of moisture and fertilizers. An excess of clay makes a heavy soil retentive of moisture and fertilizers, capable of giving a firm foothold to plants, but cold, impermeable and difficult to till. Where humus predominates the soil is often *sour* from carbonic and other acids, and is usually deficient in some of the elements of plant food. From the physical standpoint a good soil contains from sixty to eighty-five per cent. of sand, from ten to thirty of clay and iron oxid, and from five to ten of humus. As, however, the physical condition of a soil depends partly on rainfall and temperature, these must be considered along with composition.

From the chemical standpoint a soil should contain all the elements which are necessary for plant growth in a condition in which they are assimilable. What these elements are is best learned from analyses of the ashes of different plants, a short table of which is here given :

	Ash, per cent.	Potash.	Soda.	Lime.	Magnesia.	Iron.	Phosphoric Acid.	Sulfuric Acid.	Silica.	Carbonic Acid.	Chlorin.
Wheat, straw	.053	18.0	.6	4.5	..	.3	4.1	..	72.4	..	..
Wheat, grain	.013	28.5	..	1.5	12.2	.2	57.3	..	.3	..	..
Barley, "	.018	13.7	6.8	2.2	8.6	1.1	39.8	.2	27.7	..	..
Peas, "	.030	35.5	2.5	10.1	11.9	..	30.1	4.7	1.5	.5	1.3
Beets, root	.062	39.0	6.0	7.0	4.4	.5	6.0	1.6	8.0	16.1	5.1
Potatoes, tubers	.040	50.0	1.5	1.8	5.4	.5	11.3	7.1	5.6	13.4	2.9

The constituents of soils may be divided into two classes—inorganic and organic. The mineral matter due to the disintegration of rocks is composed principally of lime, magnesia, oxid of iron, alumina, potash and soda combined with silica, phosphoric, carbonic and sulfuric acids. Of these the majority are usually found in sufficient abundance, the ones which are sometimes lacking being lime, potash and phosphoric acid. The organic portion of soil is known as humus, which consists of carbon, hydrogen, oxygen and nitrogen, only the last being of value to plant life.

Potash, which is derived mainly from the decomposition of felspathic rocks like granite, exists in the soil chiefly as the soluble potassium silicate. It may constitute as much as 2 per cent., though good agricultural soils contain as little as .25 per cent. Clay soils are usually richest in potash—a fact due to the retentiveness of clay and to the common origin of clay and potash.

Phosphoric acid is found in all fertile soils, usually combined with lime. It seldom exceeds 1 per cent. even in the richest soils, and the average in good soils is probably about .2 per cent.

Lime not only affords direct food for plant life, but it also liberates potash and nitrogen held in the soil in insoluble forms. A soil containing less than 1 per cent. of lime is considered to be deficient in that particular.

Nitrogen is supplied by the decaying vegetable matter of the soil. Only as fermentation takes place is it rendered assimilable. Nitrification is brought about by a microscopic ferment, which is assisted by moisture, warmth and carbonate of lime. Very rich soils may contain as much as 1 per cent. of nitrogen, though the average of good soils is .1 or .2 per cent.

In a table on the next page the composition of a number of virgin soils is given. Soil No. 1, from the Red River valley, is particularly rich in organic matter, and consequently in nitrogen. In potash also it is much above the average, and in lime and phosphoric acid it is of fair value. Calculating for the first foot only, it contains 33,000 pounds of available nitrogen, 34,000 pounds of potash, and 9,500 pounds of phosphoric acid to the acre. An average crop of wheat is said to remove 15 pounds of phosphoric acid and 23 of potash to the acre. No. 2 is a sedentary soil derived from felspathic rocks, and consequently rich in potash, but it is poor in other respects. No. 3, which is low in lime and potash, would respond readily to fertilizers, but would be easily leached.

COMPOSITION OF VIRGIN SOILS (AIR-DRIED).

Quality.	Locality.	Water.	Organic and volatile matter.	Clay.	Sand.	Oxid of iron and alumina.	Lime.	Magnesia.	Potash.	Soda.	Phosphoric acid.	Nitrogen.
1. Black loam	Red River Valley	6.12	24.68	23.33	30.35	10.38	1.78	1.64	.97	.13	.27	.94
2. Clay loam	Restigouche, N.B	1.77	5.37	46.31	34.26	9.95	.22	.85	1.00	.10	.08	.11
3. Sandy loam	Muskoka, Ont	5.33	8.90	20.50	58.41	6.13	.08	.14	.04		.17	.28
4. Grey-black alluvial silt	Fraser River, B.C	6.66	16.39	21.55	45.77	7.70	.47	.12	.49		.27	.58
5. Sandy loam	Walkerville, Ont	1.55	6.39	11.16	72.19	8.10	.02	.66	.33		.12	.23
6. Clay loam	Sackville Marsh, N.B	8.51	5.34	63.30	11.04	10.13	.12	.33	.15		.15	.12
7. Clay loam	Joliette, Que. (surface)	2.39	7.87	45.85	32.41	8.97	.80	.84	.39	.18	.27	.21
8. Clay loam	" " (subsoil)	1.60	2.06	31.73	50.85	10.43	1.03	1.20	.43	.22	.28	.03

The fourth is a good rich soil, though a little low in lime. Nos. 5, 6 and 7 are soils of average fertility, somewhat deficient in lime.

Geological Fertilizers.—Continual cropping slowly removes from the soil the mineral ingredients on which its fertility depends. True, in good farming, a portion of these are returned in the manure, but every bushel of grain and every animal that leaves the farm carries with it some of the original phosphoric acid and potash. It is of the highest importance that these be returned to the soil in some cheap and efficacious way. A number of mineral substances are found, which either native or after chemical treatment are available for this purpose.

Apatite, the geological occurrence of which has been described in an earlier chapter, is an important source of phosphoric acid. Treated with sulfuric acid it is partially changed to a soluble phosphate. Commercial superphosphates are a mixture of calcium sulfate, calcium phosphate and calcium acid phosphate, the last of which is the valuable ingredient because of its solubility. Phosphates are especially useful as a top dressing for root crops. In connection with nitrogenous fertilizers they are also a benefit to cereals. Guano and green-sand marls are other sources of phosphoric acid, which, however, are not found in Canada.

Nitrogen, the essential fertilizer of the cereals, may be obtained from three sources. Chemical compounds, such as nitrate of soda and sulfate of ammonia, are very useful because of their solubility, but they are expensive. The first occurs as Chili saltpetre, the second is a by-product in the manufacture of coal gas. A second source is the nitrogen of the air, which can be assimilated only by leguminous plants like clover and pease. If these are ploughed under while green, a store of nitrogen is laid up for future crops. A third source is the semi-decomposed vegetable matter of muck, leaf-mould and peat. The nitrogen of these is converted into assimilable forms by fermentation, a process which is aided by composting the material with barnyard manure. These mucks and peats are widely distributed through the whole Dominion. Many analyses are given in the reports of the Experimental Farms, the average number of pounds of nitrogen to the ton being thirty-eight.

There is unfortunately no mineral source of potash in Canada. The only available supply is that stored in our forests. Wood ashes, which contain from seven to twelve per cent. of potash, are the mineral constituents which the trees by a life-long process have taken from the soil. As they also contain considerable quantities of lime, phosphoric acid and other inorganic plant food, they are among the most valuable of fertilizers. To continue to export them, as in the past, is suicidal.

Lime may be supplied from several sources. Ground gypsum or landplaster is valuable not only as food, but for liberating potash and absorbing ammonia. The crude gypsum is widely distributed, and in the manufacture of superphosphates calcium sulfate is made as a by-product. Ordinary quick-lime, besides affording nourishment makes clay soils lighter and sweetens damp and peaty ones. Marl is another source of lime very widely distributed, acting like quick-lime but more slowly. It is essentially carbonate of calcium, with more or less clay. Mussel mud is much used on Prince Edward Island, where lime is frequently lacking.

A number of other fertilizers not directly of mineral origin may be passed over. Those briefly enumerated here may, by judicious use, be made to increase the productive capacity of the soil. Questions of expense compared with returns received, of the mode and amount of application, etc., belong to agriculture rather than to economic geology, and cannot be discussed here.

LITERATURE.—Origin of Soils: Geikie, "Geology"; Shaler, Rep. U.S. Geol. Sur., XII. 1892. Analyses of Soils and Fertilizers: Shutt, Annual Reports of Experimental Farm, Ottawa.

APPENDIX.

SUMMARY OF THE MINERAL PRODUCTION OF CANADA IN 1894 AND 1895.

Product.	Calendar Years.			
	1894.		1895.	
	Quantity.	Value.	Quantity.	Value.
Metallic.				
Copper (fine in ore, etc.)..lbs.	7,737,016	$735,017	8,789,162	$949,229
Gold ....oz.	58,058	1,042,055	92,448	1,910,900
Iron ore.............tons.	109,991	226,611	102,797	238,070
Lead (fine in ore, etc.).lbs.	5,703,222	185,355	23,075,892	749,966
Mercury .... "	....	....	....	2,343
Nickel (fine, in ore, etc.). "	4,907,430	1,870,958	3,888,525	1,360,984
Platinum ....oz.	....	950	....	3,800
Silver (fine, in ore, etc) "	847,697	534,049	1,775,683	1,158,633
Total metallic.........	....	$4,594,995	....	$6,373,925
Non-Metallic.				
Arsenic (white)........tons.	7	$420	....	....
Asbestos............ "	7,630	420,825	8,756	$368,175
Chromite .... "	1,000	20,000	3,177	41,301
Coal................. "	3,867,742	8,499,141	3,513,496	7,727,446
Coke................ "	58,044	148,551	53,356	143,047
Fireclay .... "	539	2,167	1,329	3,492
Grindstones .... "	3,757	32,717	3,475	31,932
Gypsum .... "	223,631	202,031	226,178	202,608
Limestone for flux.... "	35,101	34,347	34,579	32,916
Lithographic stone.... "	180	30,000	....	2,000
Manganese ore........ "	74	4,180	125	8,464
Mica	....	45,581	....	65,000
Mineral pigments—				
Baryta ....tons.	1,081	2,830	....	....
Ochres .... "	611	8,690	1,339	14,600
Mineral water ....galls.	561,460	100,040	739,382	126,048
Moulding sand........tons.	6,214	12,428	6,765	13,530

SUMMARY OF THE MINERAL PRODUCTION OF CANADA.—*Continued.*

PRODUCT.	Calendar Years.			
	1894.		1895.	
	Quantity.	Value.	Quantity.	Value.
Non-metallic.				
Natural gas	....	$313,754	....	$423,032
Petroleum .... brls.	829,104	835,322	728,665	1,090,520
Phosphate (apatite) .....tons.	7,290	43,740	1,822	9,565
Precious stones..............	....	1,500	....	....
Pyrites ....tons.	40,527	121,581	34,198	102,594
Quartz......................	....	....	....	....
Salt ....tons.	57,199	170,687	52,376	160,455
Soapstone .... "	916	1,640	475	2,138
Whiting ....brls.	500	750	....	....
Structural materials and clay products—				
Bricks.............M.	....	1,800,000	*308,836	1,670,000
Building stone..........	....	1,200,000	....	*1,095,000
Cement, natural ....brls.	} 108,142	144,637	128,294	173,675
do Portland .... "				
Flagstones ....sq. ft.	152,700	5,298	80,005	6,687
Granite ....tons.	16,392	109,936	19,238	84,838
Lime .... bush.	....	*900,000	*5,225,000	700,000
Marble ....tons.	....	....	200	2,000
Pottery	....	162,144	....	151,588
Roofing cement ....tons.	815	3,978	....	3,153
Sands and gravels, exports .... "	324,656	86,940	277,162	118,359
Sewer pipe...............	....	250,325	....	257,045
Slate ....tons.	....	75,550	....	58,900
Terra cotta.......... "	....	65,600	....	195,123
Tiles ....M.	....	200,000	*19,200	210,000
Total non-metallic	....	$16,057,330	....	$15,295,231
do metallic........	....	4,594,995	....	6,373,925
Estimated value of mineral products not returned..............	....	297,675	....	330,844
Total	....	$20,950,000	....	$22,000,000

*Partly estimated.

APPENDIX.

TOTAL PRODUCTION.

1887	$12,500,000
1888	13,500,000
1889	14,500,000
1890	18,000,000
1891	20,500,000
1892	19,500,000
1893	19,250,000
1894	20,950,000
1895	22,000,000
1896	23,600,000*

Total for ten years $184,300,000
*Partly estimated.

The following table, compiled from figures published in Rothwell's "Mineral Industry," shows the relative standing in 1895 of the countries named in the production of some of the important minerals. In several cases countries are surpassed by others not named in the table:

	Asbestos.	Coal.	Copper.	Gold.	Iron.	Lead.	Nickel.	Petroleum.	Salt.	Silver.
Austria	..	5	8	7	7	8	4	3	7	7
Australia	..	8	5	2	..	5	..	..	..	3
Belgium	..	6	..	..	8	7	..	..	..	..
Canada	1	9	7	6	9	10	1	4	8	8
France	..	4	..	8	5	9	..	6	5	6
Germany	..	3	3	5	3	3	3	5	4	4
Great Britain	..	1	9	..	2	6	..	..	1	..
Mexico	..	..	4	4	..	4	..	..	..	1
Russia	..	7	6	3	6	11	..	2	3	9
Spain	..	10	2	..	4	1	..	..	6	5
United States	2	2	1	1	1	2	2	1	2	2

www.ingramcontent.com/pod-product-compliance
Lightning Source LLC
Chambersburg PA
CBHW020915230426
43666CB00008B/1463